APPLIED MECHANICS
WITH SOLIDWORKS

APPLIED MECHANICS WITH SOLIDWORKS

Godfrey Onwubolu

Sheridan Institute of Technology & Advanced Learning, Canada

Imperial College Press

Published by

Imperial College Press
57 Shelton Street
Covent Garden
London WC2H 9HE

Distributed by

World Scientific Publishing Co. Pte. Ltd.

5 Toh Tuck Link, Singapore 596224

USA office: 27 Warren Street, Suite 401-402, Hackensack, NJ 07601

UK office: 57 Shelton Street, Covent Garden, London WC2H 9HE

Library of Congress Cataloging-in-Publication Data
Onwubolu, Godfrey C., author.
   Applied mechanics with SolidWorks / Godfrey Onwubolu, Sheridan Institute of Technology &
Advanced Learning, Canada.
      pages cm
   Includes bibliographical references and index.
   ISBN 978-1-78326-380-6 (hardcover : alk. paper)
   1. Statics--Data processing. 2. Dynamics--Data processing. 3. Deformations (Mechanics)
4. SolidWorks. I. Title.
   TA351.O59 2014
   620.100285'53--dc23

                                                                                      2014038897

British Library Cataloguing-in-Publication Data
A catalogue record for this book is available from the British Library.

Typeset by Stallion Press
Email: enquiries@stallionpress.com

Printed in Singapore

This book is dedicated entirely to God who is in control of the mechanics behind all creation. I owe Him all that I have because all that I have comes from Him.

About the Author

Dr. Godfrey Onwubolu currently teaches Applied Mechanics, Mechanical Design, Engineering Graphics and SolidWorks, amongst other related courses. He holds a BEng degree in mechanical engineering and both an MSc and PhD from Aston University, Birmingham, England, where he first developed a geometric modeling system for his graduate studies. He worked in a number of manufacturing companies in the West Midlands, England and he was a professor of manufacturing engineering at the University of the South Pacific, teaching courses in engineering design and manufacturing for several years.

He has published several books with international publishing companies including Imperial College Press, Elsevier, and Springer-Verlag, and has published over 120 articles in international journals. He is an active Senior Member of both the American Society of Manufacturing Engineers (ASMfgE) and the American Institute of Industrial Engineers (IIE).

Table of Contents

Part 2: Dynamics

Preface

This textbook was written to assist students in colleges and universities, designers, engineers, and professionals interested in using SolidWorks to solve practical engineering mechanics problems.

The textbook is divided into two parts. Part I covers statics while Part II covers dynamics (kinematics and kinetics). *Applied Mechanics with Solid-Works* is the first textbook to use SolidWorks as an alternative tool for solving statics and dynamics problems.

Dynamics consists of kinematics and kinetics. Kinematics is the analysis of the geometry of motion without concern for the forces causing the motion; it involves quantities such as displacement, velocity, acceleration, and time. In this study of kinematics, we will consider motion in one plane only; such motion will be one of three types: (1) Rectilinear or translational motion, (2) Circular motion, and (3) General plane motion. Kinetics is the study of motion and the forces associated with motion; it involves the determination of the motion resulting from given forces. Kinetics, the study of unbalanced forces causing motion, can be analyzed by three methods: (1) Inertia force or torque (dynamic equilibrium), (2) Work and energy, and (3) Impulse and momentum. These topics are covered in Part II of this textbook.

In applied mechanics, students should attempt to utilize *logic, experience*, and *visualization* as much as possible. One of the main challenges in solving engineering mechanics problems is the ability of students to visualize how machine members that constitute an assembly move in space. Experienced designers may be able to visualize correctly how machine members move, but this can be a considerable challenge to students who have limited industrial experience.

The motivation for writing this textbook can be traced to Winter Semester 2010, when I taught Plane Motion, a topic in dynamics, to students and asked them to sketch the loci of some specific points on the machine parts as the mechanisms moves in space. Most students could not figure out the answer. That led me to initiate the Applied Mechanics Virtual Library (AMVL) project, aimed at using SolidWorks to model animated mechanisms in a machine so that students can understand how machine parts move while in operation. The other components of the AMVL initiative include the actual analysis for motion (kinematics) and the study of motion including the effect of force (kinetics) using SolidWorks. When AMVL was used to solve problems, students could visualize how the mechanisms moved and they unanimously affirmed AMLV to be an extremely useful tool for solving dynamics problems. Consequently, the AMVL project facilitated the teaching and learning of applied mechanics very significantly and the measured outcome was that students had a much better understanding of the course. That is the primary objective of this textbook. Another motivation for writing this textbook is that there is currently no known textbook that uses SolidWorks to support the teaching of applied mechanics.

What is the basis for the success of AMVL? It is simply the fact that SolidWorks is a user-friendly, efficient, and effective computer-aided design (CAD) software for modeling (which supports visualization) and analysis of machine members. These features of SolidWorks seem not to have been explored by many users of SolidWorks. In my opinion, SolidWorks is an excellent CAD software for:

- Shape design (geometric modeling).
- Machine element design.
- Statics and dynamics analysis.
- Stress analysis based on finite element analysis (FEA).

The advantages of the AMVL SolidWorks-based architecture are as follows:

- Useful for functional design.
- Enhances visualization of motion of mechanism members.
- Useful for design and analysis of mechanisms.
- Can show areas of problems during motion.
- Applied to topics in statics.
- Applied to topics in kinematics especially plane motion.
- Applied to topics in kinetics.

This textbook is written using a hands-on approach in which students can follow the steps described in each chapter to model parts and analyze them. This textbook has a significant number of pictorial descriptions of the steps that a student should follow. This approach makes it easy for users of the textbook to work on their own as they can use the steps described as guides. Instructional support is also provided, including SolidWorks files for all models. Additionally, video files are available which allow lecturers/instructors to show the motions and results of the dynamical systems being analyzed.

All the examples in this textbook have been solved by myself and checked. The textbook can be used as a Visual Laboratory tool as well as problem-solving tool for teaching statics and dynamics.

Resources for Lecturers/Instructors

SolidWorks resources designed to support the lecturers/instructors of students using this textbook have been made available for download from the publisher's website. The files provide full access to all the SolidWorks files used in the textbook. Please note that access to these files is restricted and will require an access code (available from the publisher) before they can be downloaded.

http://www.worldscientific.com/worldscibooks/10.1142/p921-sm

Acknowledgements

Michael Arthur provided five problems in the dynamics section which I solved while we both taught the applied mechanics course. Michael Arthur and Lewis Mununga have been a great team with whom I have been teaching applied mechanics at Sheridan Institute of Technology & Applied Learning, Canada; they are both acknowledged. When I first suggested the idea of writing a book on applied mechanics using SolidWorks, Lance Sucharov showed interest in my sending him an outline of the proposed book. I would especially like to thank him and Matthew Judge (who moved on before the completion of this book project) as well as Catharina Weijman (who completed this book project) of Imperial College Press for their helpful suggestions and assistance throughout this book project. My wife, Ngozi, and our children are greatly appreciated. My wife shared some very challenging times with me throughout the process of writing this textbook. Without the role that members of my family played, this textbook project would not have succeeded.

Godfrey Onwubolu Toronto, Canada May 2014

PART 1
Statics

Chapter 1

Introduction

Objectives: When you complete this chapter you will have:

- Understanding of the different areas of applied mechanics.
- Understanding of how SolidWorks can be applied to the different areas of applied mechanics.

Applied Mechanics

As a scientific discipline, *applied mechanics* derives many of its principles and methods from the physical sciences (in particular, mechanics and classical mechanics), from mathematics and, increasingly, from computer science. As such, applied mechanics shares similar methods, theories, and topics with applied physics, applied mathematics, and computer science.

Applied mechanics, as its name suggests, bridges the gap between physical theory and its application to technology. As such, applied mechanics is used in many fields of engineering, especially mechanical engineering. In this context, it is commonly referred to as engineering mechanics. Much of modern engineering mechanics is based on Isaac Newton's laws of motion while the modern practice of their application can be traced back to Stephen Timoshenko, who is said to be the father of modern engineering mechanics.

Applied mechanics in engineering

Typically, engineering mechanics is used to analyze and predict the acceleration and deformation (both elastic and plastic) of objects under known forces (also called loads) or stresses.

When treated as an area of study within a larger engineering curriculum, *engineering mechanics* can be subdivided into:

- *Statics*, the study of non-moving bodies under known loads.
- *Dynamics*, the study of how forces affect moving bodies.

- *Mechanics of materials or strength of materials*, the study of how different materials deform under various types of stress.
- *Deformation mechanics*, the study of deformations typically in the elastic range.
- *Fluid mechanics*, the study of how fluids react to forces. Note that fluid mechanics can be further split into fluid statics and fluid dynamics, and is itself a sub-discipline of continuum mechanics. The application of fluid mechanics in engineering is called hydraulics.
- *Continuum mechanics* is a method of applying mechanics that assumes that all objects are continuous. It is contrasted by discrete mechanics and the finite element method.

In this textbook, *statics* and *dynamics* as well as some elements of *strength of materials* are covered. Although objects encountered in engineering are mainly solids or fluids, this book focuses on solids. Consequently, fluid mechanics is not covered. Deformation and continuum mechanics are specialized areas and are also not covered.

Scope of Applied Mechanics

The scope of applied mechanics covered in this textbook is shown in Figure 1.1, which embraces virtually all topics covered in standard college and university curriculum in this subject. The new approach followed in this textbook is to apply SolidWorks to the entire spectrum of Figure 1.1.

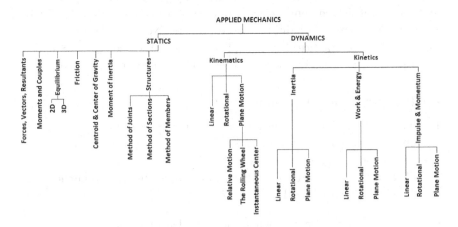

Fig. 1.1 Scope of applied mechanics.

Forces, vectors, and resultants are treated in Chapter 2; moments and couple are treated in Chapter 3 while two-dimensional (2D) equilibrium is treated in Chapter 4. In Chapter 5, structures and members are covered. It should be noted that while analytical methods distinguish truss frames, machine and members in the ways in they are handled, SolidWorks has the capability of dealing with all three sub-divisions with ease. Three-dimensional (3D) equilibrium is treated in Chapter 6; friction is treated in Chapter 7 while centroid is treated in Chapter 8 and moment of inertia is treated in Chapter 9. These topics constitute the section on statics.

The second half of the textbook covers dynamics. Kinematics: Linear and projectiles is treated in Chapter 10; while kinematics: angular motion is treated in Chapter 11. Kinematics: Plane motion of mechanisms is treated in Chapter 12. This chapter discusses the concept of the Applied Mechanics Virtual Library (AMVL) which the author introduced as a new pedagogy in teaching applied mechanics; the chapter also discusses in detail velocity and acceleration diagrams, which are the key tools applied to Chapter 13 which concludes the topic of kinematics. The following sub-section treats kinetics. Kinetics: Force and torque inertia is treated in Chapter 14; Kinetics: Work and energy is treated in Chapter 15; Kinetics: Impulse and momentum is treated in Chapter 16. All three sub-sections of kinetics cover linear, angular, and plane motions.

The Method of Problem Solution in Engineering Mechanics

The general method of problem solution and workmanship used in the discipline of engineering is used in engineering mechanics. Students should attempt to utilize *logic, experience*, and *visualization* as much as possible. One of the main challenges in solving engineering mechanics problems is the ability of students to visualize how machine members that constitute an assembly move in space. Experienced designers may be able to visualize correctly how machine members move, but this can be a considerable challenge to students who have limited industrial experience. General steps to finding solutions in engineering mechanics are:

1. Sketch the object, frame, machine member, etc., reproduced in a concise form suitable for referencing during problem solution.
2. State given information not already labeled on sketch.
3. State required information.
4. Produce the first free-body diagram (FBD) or sketch.

5. State the type of calculation to be used.
6. Calculations.
7. Produce the next FBD or sketch, if necessary.
8. State the type of calculation to be used.
9. Calculations (repeat Steps 7–9 where necessary, until final answer is obtained).

The traditional method of problem solution and workmanship in engineering mechanics requires the student and indeed a designer, engineer, or engineering technologist to be strong in *geometry* and *trigonometry*. Without a good background in these areas, it will be extremely challenging to pursue the solutions of most problems in engineering mechanics to a conclusion. The combined problem of *visualization* and *mathematics* is the one main reason most students taking engineering mechanics at college or university find it quite challenging.

Engineering Mechanics Solutions Using SolidWorks

It was stated in the preceding sub-section that correctly visualizing how machine members move, is a considerable challenge to students who have limited industrial experience. SolidWorks is a powerful CAD software that has the capability of enhancing visualization as well as providing solutions to engineering mechanics problems in a very user-friendly manner.

The outputs of statics problems are in graphical form, showing that forces and other parameters are joints and/or structural members. Throughout the book, kinematics problems are solved and answers are given graphically so that students can relate to the outputs. The most interesting part is kinematics where SolidWorks Motion Simulations produce simulation results with mechanisms actually moving. Not only is this approach useful in theoretically analyzing the dynamics of machines, it is also useful to engineers as a practice for the initial design of machine parts and mechanisms.

An Illustrative Example of AMVL

As an example of the capability of AMVL see Figures 1.2 to 1.4. The block is moved along the slot and students observe how the links move. A linear motor moves the block while the animation output is observed. The velocity diagram showing the solution to the problem is obtained. All these are done using SolidWorks.

Assembly

Fig. 1.2 Assembly model of a mechanism.

Motion

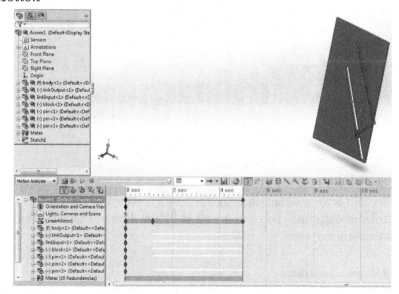

Fig. 1.3 Motion study output for the assembly model.

Analysis

The steps leading to the analysis solution is not described in this chapter; it is explained in detail in Chapter 12. The idea is to show that the assembly, motion analysis, and velocity diagram can be effectively handled using SolidWorks.

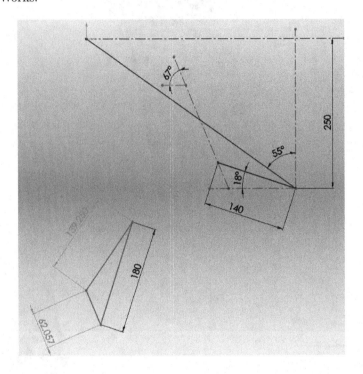

Fig. 1.4 Velocity diagram showing the solution to the problem.

Problem 1.1

A roller and lever mechanism is in the position shown in Figure 1.5. Find the horizontal distance between A and B. (Point B is at the same level as the horizontal surface.)

SolidWorks solution

Open a New SolidWorks Document.
Choose the Front Plane.

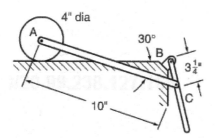

Fig. 1.5 Problem description.

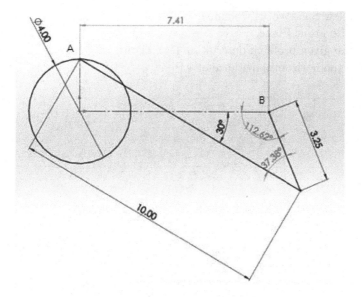

Fig. 1.6 Sketch of the given problem description.

Sketch the given problem description (See Figure 1.6).
Measure the horizontal distance of AB.
The horizontal distance between A and B = 7.41 in.

Problem 1.2

A roller and lever mechanism is in the position shown in Figure 1.7. Find the horizontal distance between A and B. (Point B is at the same level as the horizontal surface.)

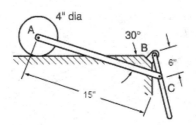

Fig. 1.7 Problem description.

SolidWorks solution

Open a New SolidWorks Document.
Choose the Front Plane.
Sketch the given problem description (see Figure 1.8).
Measure the horizontal distance of AB.

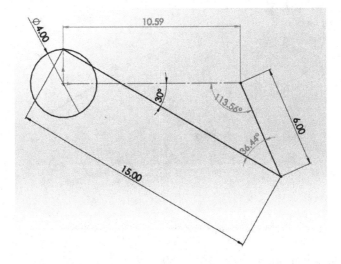

Fig. 1.8 Sketch of the given problem description.

The horizontal distance between A and B = 10.59 in.

Problem 1.3

Neglecting the pulley diameter at D, determine how far weight A drops as θ changes from 120° to 60° (see Figure 1.9).

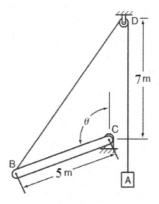

Fig. 1.9 Problem description.

SolidWorks solution

Open a New SolidWorks Document.
Choose the Front Plane.
Sketch the given problem description (see Figure 1.10).
Measure the difference between BD and $BD' = 10.44 - 6.24 = 4.2$".

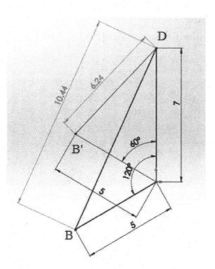

Fig. 1.10 Sketch of the given problem description.

Problem 1.4

In the mechanism shown in Figure 1.11, block C slides to the left until member BC is vertical. Determine the change in length of distance AC and also the change in movement CC'.

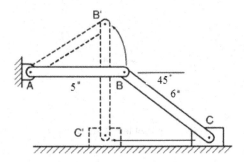

Fig. 1.11 Problem description.

SolidWorks solution

Open a New SolidWorks Document.
Choose the Front Plane.
Sketch the given problem description (See Figure 1.12).

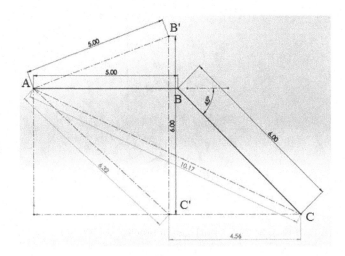

Fig. 1.12 Sketch of the given problem description.

Measure the difference between AC and $AC' = 10.17 - 6.32 = 3.85$".
Measure the difference between CC and $CC' = 4.56$".

Summary

The example shown in this chapter highlights the advantages of the AMVL framework. Students learn applied mechanics using a new paradigm: Solid-Works is used for modeling and visualization of the machine being studied (using SolidWorks' Assembly and Motion Analysis tools) and for the actual analysis (Analysis).

Exercises

P1. The length of member BC in Figure P1 is 85 mm. Determine angle ϕ.

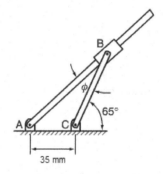

Fig. P1

P2. The toggle mechanism shown in Figure P2 moves from an initial angle $\theta = 30°$ to a final angle, $\theta = 20°$. Calculate the horizontal distance moved by point C.

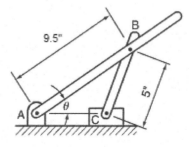

Fig. P2

P3. Lever AB is 2 m long and initially horizontal as shown in Figure P3. Determine the angle θ that lever CD must rotate to cause lever AB to rotate 30° upward.

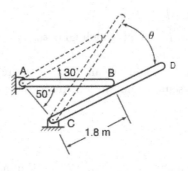

Fig. P3

Reference

Walker, K. M., *Applied Mechanics for Engineering Technologists*, 8th Edition, Prentice Hall, Upper Saddle River, NJ, 2007.

Chapter 2

Forces, Vectors, and Resultants

Objectives: When you complete this chapter you will have:

- Understanding of how SolidWorks determines the resultants of vectors that are not necessarily at right angles to each other.
- Used SolidWorks to determine resultants of several vectors by method of components.

Vectors

Vectors are quantities that require not only a magnitude, but a direction to specify them completely. Let us illustrate by first citing some examples of quantities that are not vectors. The number of gallons of gasoline in the fuel tank of your car is an example of a quantity that can be specified by a single number — it makes no sense to talk about a 'direction' associated with the amount of gasoline in a tank. Such quantities, which can be specified by giving a single number (in appropriate units), are called *scalars*. Other examples of scalar quantities include the temperature, your weight, or the population of a country; these are scalars because they are completely defined by a single number (with appropriate units).

However, consider a velocity. If we say that a car is moving 70 km/hour, we have not completely specified its motion, because we have not specified the *direction* that it is going. Thus, velocity is an example of a vector quantity. A vector generally requires more than one number to specify it; in this example we could give the magnitude of the velocity (70 km/hour), a compass heading to specify the direction (say 30° from north), and a number giving the vertical angle with respect to the Earth's surface (0°; except in chase scenes from action movies!).

There are a number of examples of vectors in static mechanics. All forces are represented by vectors. Additionally, distance, velocity, and acceleration are vectors.

Force Types, Characteristics, and Units

Applied force: This is a very real and noticeable force applied directly to an object, such as in the case of the force that you would apply to a book.

Weight or non-applied force: The weight of a book on a table is a non-applied force. In this case, the weight is the concentrated force acting at the centre of gravity of the book

Distributed load: The same weight of a book on the table could be shown as a distributed load, consisting of many smaller forces distributed over the entire surface of the book.

Internal and external forces: An internal force is a force inside a structure, and an external force acts outside the structure. Consider the FBD of a member involving loads and reactions in a plane (see Figure 2.1). The beam AB has external forces, P_1 and P_2. The beam experiences internal forces, being compressive on the top and in tension at the lower part. There are external reactions A_x, A_y, and B.

Resultants

When vectors are added, their direction must be considered. For example, if a bicycle rider moves 4 km east and then 3 km north, the resultant is 5 km north-east (see Figure 2.2).

Vector Addition: Graphical

As discussed, the sum of two or more vectors is a resultant. Graphical vector addition requires the drawing of the vector to some scale in their given direction. The resultant can then be measured or scaled from the drawing.

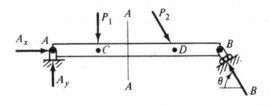

Fig. 2.1 Free-body diagram of a member involving loads and reactions in a plane.

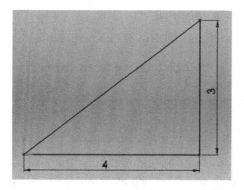

Fig. 2.2 Resultant of vectors.

There are three known methods or rules of vector addition:

Triangle: Used when forces form a right-angle triangle or any other triangle (Figure 2.3(a) and 2.3(b))

Parallelogram: The diagonal of the parallelogram formed represent the resultant (Figure 2.3(c))

Vector polygon: A continuation of the triangle rule to accommodate more than two forces (Figure 2.3(d))

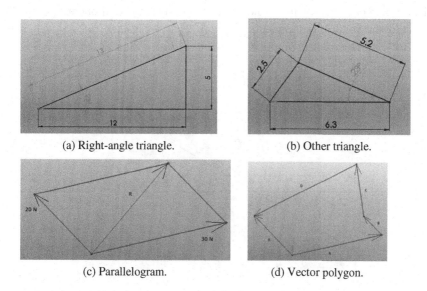

(a) Right-angle triangle. (b) Other triangle.

(c) Parallelogram. (d) Vector polygon.

Fig. 2.3 Vector addition.

Vector Addition: Analytical

Analytical vector addition consists of two main methods:

1. Construction of a triangle and use of the cosine law or other simple trigonometric functions.
2. Addition of the components of vector.

Components

Resolution of a vector into its component is the reverse of adding to get the resultant. A single force inclined at an angle can be broken up into two separate forces — this is known as resolution of a force into its components.

As an illustration, a force of 200 N inclined at 30° to the horizontal shown in Figure 2.4 is represented by a vector of 20 units, resulting in horizontal component of 173 N, east, and a vertical component of 100 N, north.

Vector Addition: Components

To add vectors analytically using the method of components requires the following steps:

1. Resolve each vector into a horizontal and vertical component.
2. Add the vertical components, $R_y = \sum F_y$.
3. Add the horizontal components, $R_x = \sum F_x$.

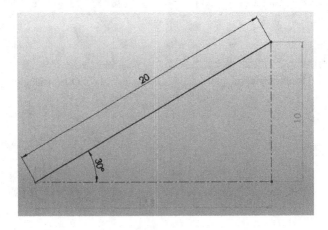

Fig. 2.4 Components of a force.

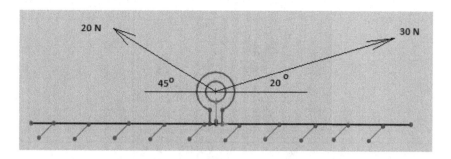

Fig. 2.5 Force systems.

4. Combine the vertical and horizontal components to get a single resultant vector, $R = \sqrt{R_x^2 + R_y^2}$.

Problem 2.1

Forces of 20 N and 30 N are pulling on a ring (Figure 2.5). Determine the resultant using the triangle rule.

SolidWorks solution

1. Open a New SolidWorks Part Document.
2. Select the Front Plane.
3. Sketch a line parallel to the 30 N force at an angle of 30° to the horizontal.
4. At the end of the first sketch, sketch another line parallel to the 20 N force at an angle of 45° to the horizontal.
5. Sketch a line from the end of the first line to the end of the second line (see Figure 2.6). (This last line is the resultant of the two forces with a value of 28 N.)

Problem 2.2

Determine the resultant of the vectors shown in Figure 2.7.

SolidWorks solution

1. Open a New SolidWorks Part Document.
2. Select the Front Plane.
3. Sketch a line parallel to the 180 N force that is horizontal.

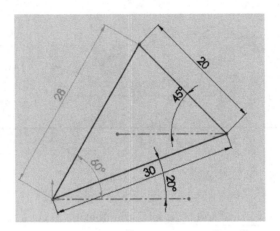

Fig. 2.6 Resultant of two forces.

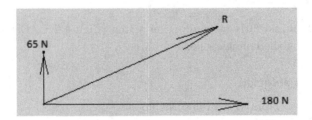

Fig. 2.7 Force system.

4. At the end of the first sketch, sketch another line parallel to the 65 N force at an angle of 90° to the horizontal.
5. Sketch a line from the end of the first line to the end of the second line (see Figure 2.8).

   (This last line is the resultant of the two forces; with a value of 191.38 N, at an angle of 20° to the horizontal.)

Problem 2.3

Solve for the resultant of the force system shown in Figure 2.9 acting on point A.

SolidWorks solution

1. Open a New SolidWorks Part Document.

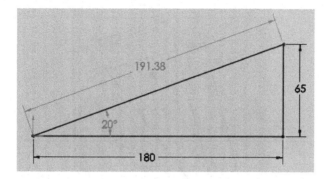

Fig. 2.8 Resultant of two forces.

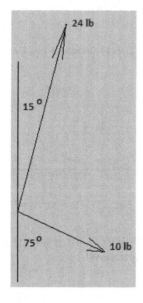

Fig. 2.9 Force system.

2. Select the Front Plane.
3. Sketch a line parallel to the 24 lb force that is inclined 15° to the vertical.
4. At the end of the first line, sketch another line parallel to the 10 lb force at an angle of 75° to the vertical.
5. Sketch a line from the end of the first line to the end of the second line (see Figure 2.10).

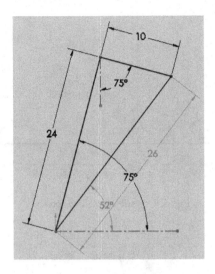

Fig. 2.10 Resultant of two forces.

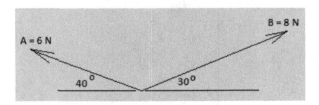

Fig. 2.11 Force system.

(This last line is the resultant of the two forces; with a value of 26 lb, at an angle of 52° to the horizontal.)

Problem 2.4

Find the resultant in Figure 2.11.

SolidWorks solution

1. Open a New SolidWorks Part Document.
2. Select the Front Plane.
3. Sketch a line parallel to the 8 N force that is inclined 30° to the horizontal.

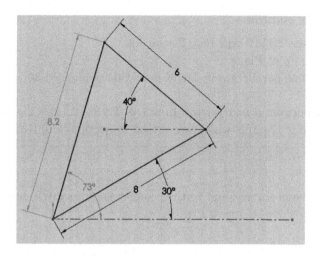

Fig. 2.12 Resultant of two forces.

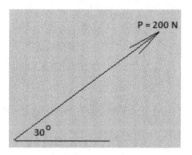

Fig. 2.13 Force system.

4. At the end of the first line, sketch another line parallel to the 6 N force at an angle of 40° to the horizontal.
5. Sketch a line from the end of the first line to the end of the second line (see Figure 2.12).

   (This last line is the resultant of the two forces; with a value of 8.2 N, at an angle of 73° to the horizontal.)

Problem 2.5

Determine the horizontal and vertical components of P (Figure 2.13).

SolidWorks solution

1. Open a New SolidWorks Part Document.
2. Select the Front Plane.
3. Sketch a line parallel to the 200 N force that is inclined 30° to the horizontal.
4. Project horizontal and vertical lines (see Figure 2.14). (The horizontal projection is 173.21 N while the vertical projection is 100 N.)

Problem 2.6

Determine the horizontal and vertical components when the direction of P is shown as a slope (Figure 2.15).

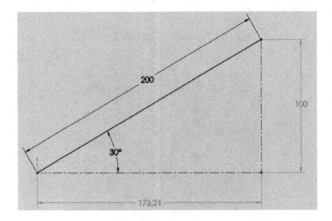

Fig. 2.14 Components of force.

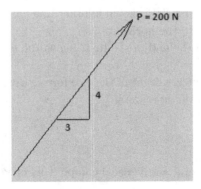

Fig. 2.15 Force system.

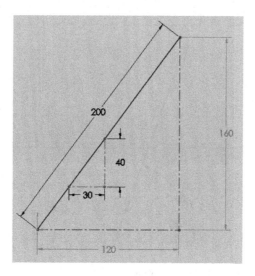

Fig. 2.16 Components of force.

SolidWorks solution

1. Open a New SolidWorks Part Document.
2. Select the Front Plane.
3. Sketch a line parallel to the 200-N force that is inclined 3 units horizontal, 4 units vertical.
4. Project horizontal and vertical lines (see Figure 2.16). (The horizontal projection is 120 N while the vertical projection is 160 N.)

Problem 2.7

Find the components of force Q for the axis system of A and B as shown in Figure 2.17.

SolidWorks solution

1. Open a New SolidWorks Part Document.
2. Select the Front Plane.
3. Sketch a line parallel to the 100 lb force (Q) that is inclined 70° to the horizontal.
4. Sketch a line, 40° to the horizontal (A); sketch another line 60° to the horizontal (B).
5. Project line Q to lines A and B respectively (see Figure 2.18). (The projection of A is 77.79 lb while the projection of B is 50.77 lb.)

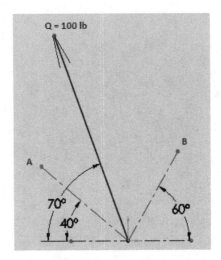

Fig. 2.17 Force system.

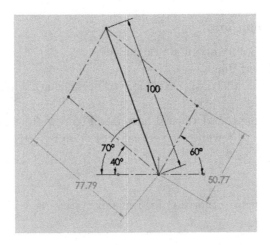

Fig. 2.18 Components of force.

Problem 2.8

Find the resultants of forces P and Q as shown in Figure 2.19.

SolidWorks solution

1. Open a New SolidWorks Part Document.

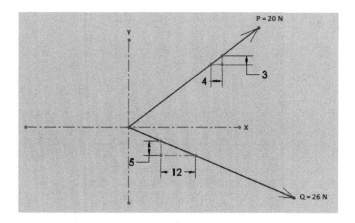

Fig. 2.19 Force system.

2. Select the Front Plane.
3. Sketch a line parallel to the 20 N force (P) that is inclined 3 units horizontal, 4 units vertical.
4. Sketch a line parallel to the 26 N force (Q) that is inclined 12 units horizontal, 5 units vertical.
5. At the end of line P, sketch a line equal and parallel to line Q (call this T).
6. Sketch a line from the end of the first line to T (see Figure 2.20). (The resultant OT is 40.05 N.)

Problem 2.9

Find the resultants of the forces shown in Figure 2.21.

SolidWorks solution

1. Open a New SolidWorks Part Document.
2. Select the Front Plane.
3. Sketch a line parallel to the 6 kips force that is inclined 0.2 unit horizontal, 0.1 unit vertical.
4. Sketch a line parallel to the 6.8 kips force that is inclined 0.8 unit horizontal; 1.5 unit vertical.
5. Sketch a line parallel to the 10 kips force that is inclined 0.7 unit horizontal; 0.4 unit vertical (see Figure 2.22).

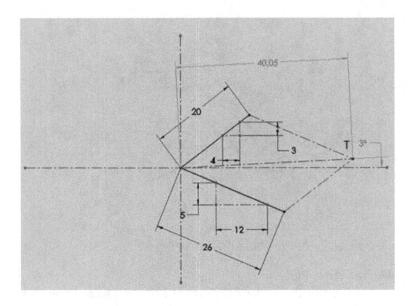

Fig. 2.20 Components of force.

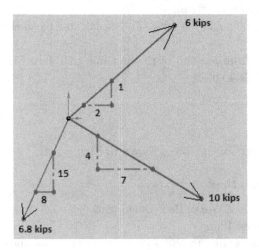

Fig. 2.21 Force system.

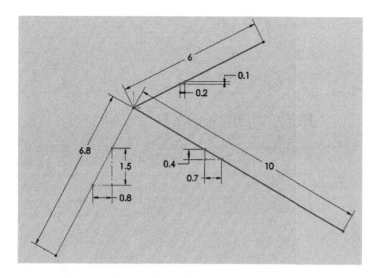

Fig. 2.22 Copy of system of forces.

6. At the end of line OB, sketch a line equal and parallel to line OA (call this P).
7. Sketch OP (see Figure 2.23; OP is resultant of OB and OA).
8. At the end of line OP, sketch a line equal and parallel to line OC (call this Q).
9. Sketch OQ (see Figure 2.23; OQ is resultant of OP and OC). (This last line OQ is the resultant of the three forces with a value of 13.65 lb.)

Problem 2.10

The tension in cable BC attached to a wall AC and a beam AB at each end is 145 lb Forces are applied to the end of the beam as shown in Figure 2.24. Determine the resultant of the forces exerted at point B of beam AB.

SolidWorks solution

Treat as a 3-force system (145 lb along BC, 156 lb, and 100 lb meeting at joint B)

1. Open a New SolidWorks Part Document.

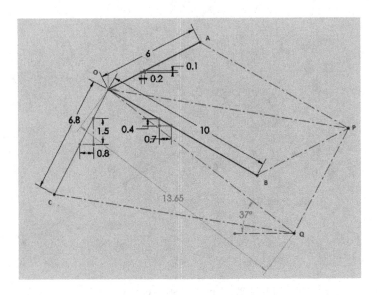

Fig. 2.23 Resultant of forces.

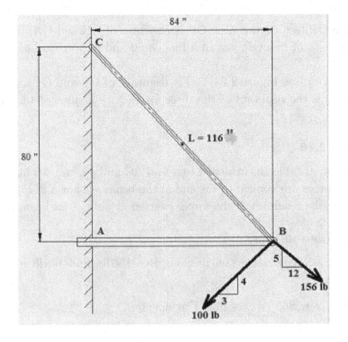

Fig. 2.24 Force system.

2. Select the Front Plane.
3. Sketch a line parallel to the 145 lb force (OA) that is inclined 8 unit horizontal, 8.4 unit vertical.
4. Sketch a line parallel to the 156 lb force (OC) that is inclined 12 unit horizontal, 5 unit vertical.
5. Sketch a line parallel to the 100 lb force (OB) that is inclined 4 unit horizontal, 4 unit vertical (see Figure 2.25).
6. At the end of line OB, sketch a line equal and parallel to and in the direction of line OA (call this end of line, D).
7. At the end of line BD, sketch a line equal and parallel to and in the direction of line OC (call this end of line, E).
8. Sketch OE (see Figure 2.23; OE is resultant of OA, OB and OC). (This last line OE is the resultant of the three forces with a value of 45.18 lb.)

Problem 2.11

A block on an inclined plane has forces shown in Figure 2.26 acting on it; the angle $\theta = 35°$. Determine the resultant force on the block.

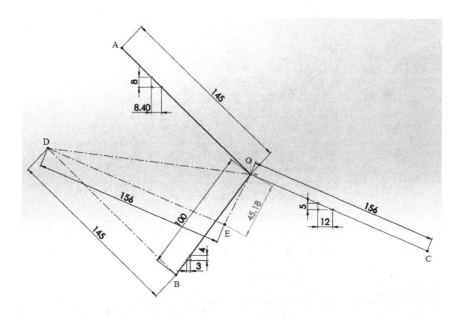

Fig. 2.25 Resultant of forces.

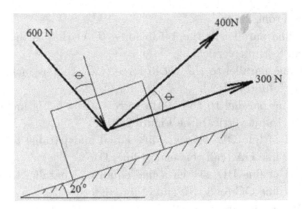

Fig. 2.26 Force system.

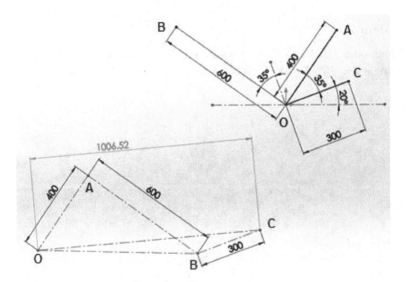

Fig. 2.27 Resultant of forces.

SolidWorks solution

Treat as a 3-force system (600 N, 400 N, and 300 N).

1. Open a New SolidWorks Part Document.
2. Select the Front Plane.
3. Sketch a line parallel to the 400 N force (OA) that is inclined 55° (20+35) to the horizontal.

4. Sketch a line parallel to the 600 N force (OB) that is inclined 25° to the normal of the inclined plane.
5. Sketch a line parallel to the 300 N force (OC) that is along the inclined plane (see Figure 2.27).
6. At the end of line OA, sketch a line equal and parallel to and in the direction of line OB (call this end of line, B).
7. At the end of line AB, sketch a line equal and parallel to and in the direction of line OC (call this end of line, C).
8. Sketch OC (see Figure 2.23; OC is resultant of OA, OB and OC). (This last line OC is the resultant of the three forces with a value of 1006.52 N.)

Summary

The features of vectors have been discussed. A number of methods for their resultants have been examined as well as methods for finding their components. The most general method is that of adding vectors analytically using the method of components. Note that CAD software such as SolidWorks, Inventor, etc. are very effective for carrying out vector addition tasks. Finding components and resultants of vectors is very easy: Simply add vectors end-to-end while maintaining their directions in order to determine the resultants; project vectors in order to determine their components. It is that easy and fast.

Exercises

P1. Determine the resultant of the forces shown in Figure P1.

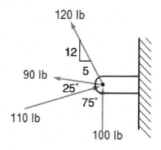

Fig. P1

P2. Determine the resultant of the forces shown in Figure P2.

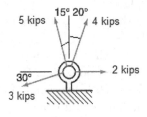

Fig. P2

P3. Determine the resultant of the forces shown in Figure P3.

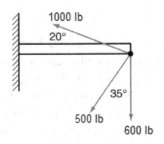

Fig. P3

P4. Determine the resultant of the forces shown in Figure P4.

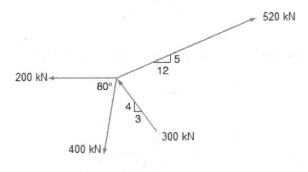

Fig. P4

Reference

Walker, K. M., *Applied Mechanics for Engineering Technologists*, 8th Edition, Prentice Hall, Upper Saddle River, NJ, 2007.

Chapter 3

Moments and Couples

Objectives: When you complete this chapter you will have:

- Understanding of how SolidWorks can be applied to determining moments and couples.
- Used SolidWorks to determine moments and couples of mechanical elements carrying loads.

Moment of a Force

Moment is the same as a couple, which is the measure of the turning effect, or torque, produced by a *force*, acting on a body. Examples include pushing a revolving door while walking through it, tightening of a nut with a wrench, turning a steering wheel of a car, unscrewing a fluorescent bulb from its fitting, etc. If a force is acting some distance away from a point, it causes a twisting action about the point. This twisting action, or torque, is called a moment. The magnitude of the moment depends upon both the size of the force and the perpendicular distance from the force to the point. Mathematically, we can define the moment as:

Moment = force × the perpendicular distance between the axis and the
line of action of the force.

Moments are vector quantities; therefore, their direction must be specified. Figure 3.1 shows the force F and the perpendicular distance d from O for determining moment/torque.

Couples

A *couple* consists of *two equal forces*, acting in *opposite directions* and *separated by a perpendicular distance*. When you turn your steering wheel with both hands, equal forces are applied and these are separated by the diameter

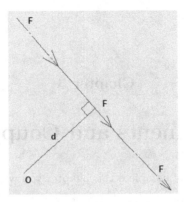

Fig. 3.1 Force and perpendicular distance.

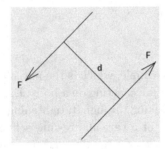

Fig. 3.2 Force and perpendicular distance.

of the steering wheel. Figure 3.2 shows the force F and the perpendicular distance d for determining the couple.

Summary

1. Moment = force × perpendicular distance, where perpendicular distance is the shortest possible distance between the line of action of the force and the point about which you are taking moments.
2. The final answer of a moment must have the direction shown.
3. When writing moment equations, we use clockwise as negative and counter-clockwise as positive.
4. There should be no force alone in a moment equation. Check that each force has been multiplied by its perpendicular distance.
5. A couple has the same moment about any point.

Rationale for Using SolidWorks to Determine Moments and Couples

Why should SolidWorks be used to determine moments and couples when these can be easily determined using analytical methods? There is no doubt that many resources are utilized when using SolidWorks and there must therefore be good reasons for using it to determine moments and couples. It turns out that when a model of interest has been created and Solid-Works Simulation is invoked not only are moments and couples outputs but there are several other outputs which can be obtained as results. This is the main advantage of using SolidWorks for this exercise. We will now consider some problems and discuss the SolidWorks methodology for their solutions.

Problem 3.1

Determine the moment or torque tightening the pipe in Figure 3.3.

SolidWorks solution

1. Open a **New SolidWorks Part** Document.
2. Select **Front Plane.**
3. Be in **Sketch** mode.
4. Create the **Sketch1** shown in Figure 3.4.
5. Exit Sketch mode.

Create structural element(s)

1. Click **Insert > Weldments > Structural Member.** (The Structural Member PropertyManager automatically appears; see Figure 3.5.)

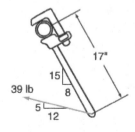

Fig. 3.3 Pipe and wrench system.

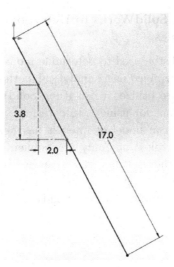

Fig. 3.4 Sketch1.

Fig. 3.5 Structural member definitions.

For Selections, choose the following.

2. **Standard: ansi inch.**
3. **Type: rectangular tube.**
4. **Size: 3 × 2 × 0.25** (this is modified; see structural member).
5. **Groups:** Select **Sketch** as **Group1.**
6. Click **OK** (see Figure 3.6 for the model required for analysis).

Analysis

1. Click **Add-Ins.**
2. Select **SolidWorks Simulation** (see Figure 3.7).
3. Click **Simulation > Study** [1] (see Figure 3.8).
   (Study PropertyManager is automatically displayed.)
4. Give Name as **Pipe Wrench** (see Figure 3.9).
5. Click **OK.**

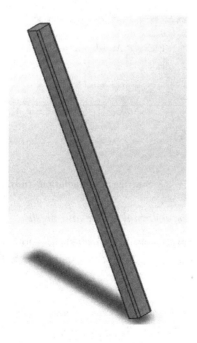

Fig. 3.6 Structural member model.

Fig. 3.7 Add-Ins interface.

Fig. 3.8 Starting the simulation study.

Apply material, fixtures, external loads, and mesh

Material, fixtures, external loads, and mesh are now applied to the model using the CommandManager interface or using the SolidWorks Simulation Manager shown in Figure 3.10.

Apply material

1. Select **Apply Material** [1] (see Figure 3.10).
2. Select **English (IPS)**; **Apply** and **Close** (see Figure 3.11).

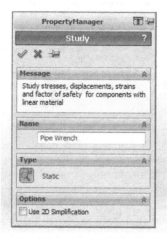

Fig. 3.9 Naming the study.

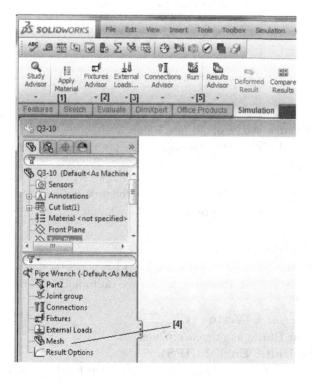

Fig. 3.10 Steps in the Simulation process.

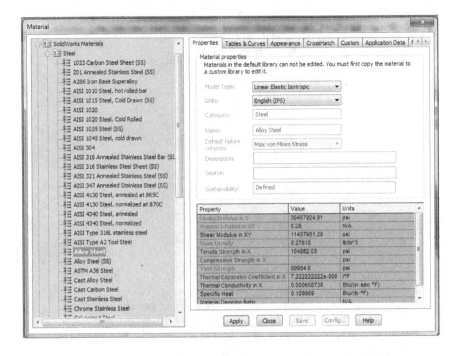

Fig. 3.11 Material selection.

Apply fixtures

1. Select **Fixture Advisor** [2] (see Figure 3.10).
2. Select **Fixed Geometry** and apply the *top node* (see Figure 3.12).
3. Click **OK**.

Since the pipe is held by the wrench at the upper joint, that joint should be fixed while the load is applied at the lower joint.

Apply external loads

(1) Select **External loads** [3] (see Figure 3.10).
    (The Force/Torque PropertyManager is automatically displayed; see Figure 3.13)
2. Select *lower node* [**Vertex ⟨1⟩**].
3. Select **Top Plane** as reference plane.
4. Select for **Units: English (IPS)**.
5. Apply *horizontal force in x-direction* relative to the *Top Plane*, with value = **36 lb**.

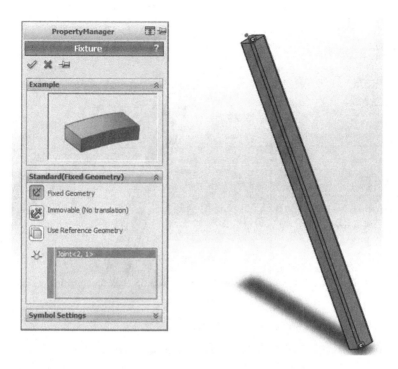

Fig. 3.12 Fixed joint (upper joint).

6. Apply *normal force* relative to the *Top Plane*, with value = **15 lb**.
7. Click **OK** to apply.

First, the inclined load is resolved into x-component (15 lb) and y-component (36 lb). The selections need some explanations. By selecting the Top Plane as reference plane, a normal force to the Top Plane will point in the y-direction (down or up). The direction can be corrected using the 'Reverse direction' check box. Direction1 corresponds to x-axis, and this direction is used for the 36 lb load.

Apply mesh and run

1. Right-click **Mesh** [4] (see Figure 3.10).
2. Select **Run** [5] (see Figure 3.10).

Viewing results

1. Right-click **Results** [1] (see Figure 3.14).

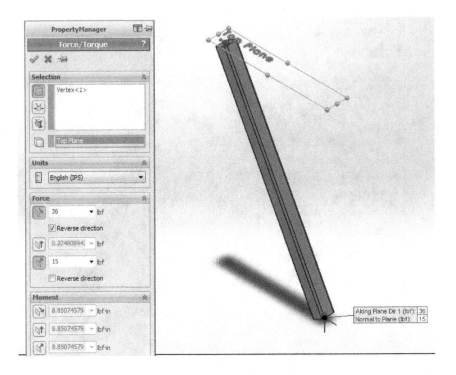

Fig. 3.13 External load applied.

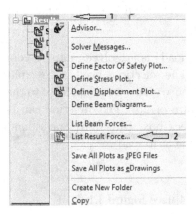

Fig. 3.14 Selecting result options.

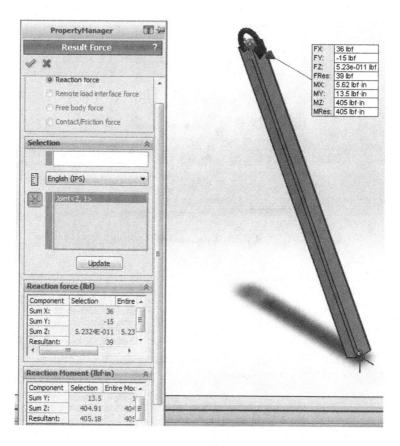

Fig. 3.15 Result Force PropertyManager.

2. Select **List Result Force ...** [2] (see Figure 3.14).
   (The **Result Force PropertyManager** is automatically displayed (see Figure 3.15).)
3. Select as **Unit, English (IPS)**.
4. Select as **Joint**, the top joint (**Joint < 2, 1 >**) (see Figure 3.15).
5. Click **Update** to show the solution.

As already mentioned, several other result options are available and moment is one of them. If the user is interested in viewing other results, this is done via **Results > Stress1 > Settings** (see Figure 3.16).

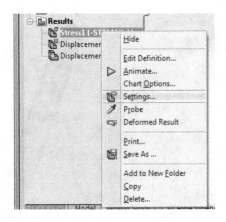

Fig. 3.16 Route to access other results.

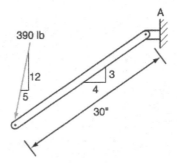

Fig. 3.17 Lever and load at its end.

Problem 3.2

Determine the moment about point A of the lever shown in Figure 3.17.

SolidWorks solution

1. Open a **New SolidWorks Part** Document.
2. Select **Front Plane**.
3. Be in **Sketch** mode.
4. Create the **Sketch1** shown in Figure 3.18.
5. **Exit Sketch** mode.

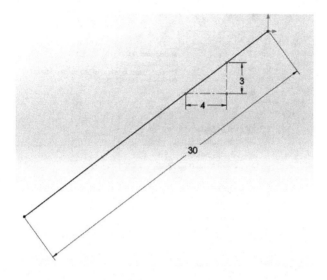

Fig. 3.18 Sketch1.

Fig. 3.19 Create structural member(s).

Fig. 3.20 Selecting result options.

Create structural element(s)

1. Click **Insert** > **Weldments** > **Structural Member** (See [1] and [2] respectively in Figure 3.19).
   (The Structural Member PropertyManager automatically appears; see Figure 3.19.)
2. Right-click **Results** folder [1] (see Figure 3.20).
3. Select **List Result Force...** [2] (see Figure 3.20).
   (The Result Force PropertyManager is automatically displayed; see Figure 3.21.)

Summary

Using SolidWorks Simulation to determine the moment about point commits expensive resources to solve a trivial problem. However, the advantage of using SolidWorks Simulation to determine the moment is that several other results are made available such as axial stress, shear stress, etc., as shown in Figure 3.22.

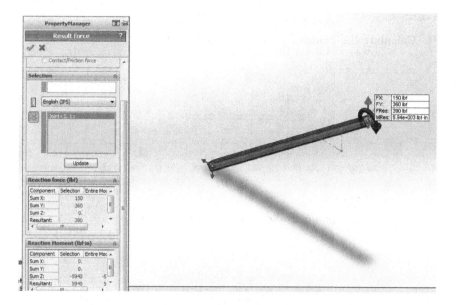

Fig. 3.21 Result Force PropertyManager.

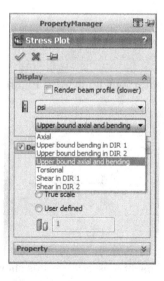

Fig. 3.22 Several other results (apart from moment) available.

Exercises

P1. Calculate the moment about point A in Figure P1.

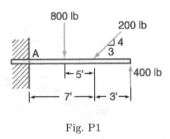

Fig. P1

P2. Calculate the moment about point A in Figure P2.

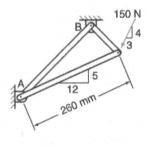

Fig. P2

P3. Calculate the moment about point A in Figure P3.

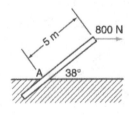

Fig. P3

P4. The boom AD in a machine is extended as shown in Figure P4. Using the following dimensions in feet: AC = 3; AB = 6; BC = 4; BD = 5, determine:

(a) The moment about A due to the applied load, W = 1000 lb.

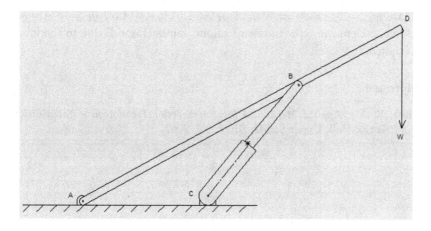

Fig. P4

(b) The cylinder force pushing in the same direction as the cylinder at B, if its moment is equal to 80% the moment of the 1000 lb force about point A.

P5. The truss in Figure P5 is loaded with $W_1 = 475\,lb$, $W_2 = 140\,lb$ and $W_3 = 250\,lb$ as shown. The horizontal and vertical measurements

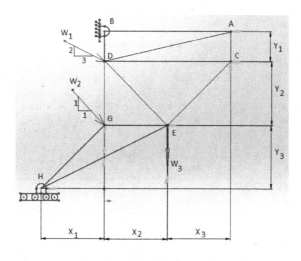

Fig. P5

are $x_1 = x_2 = x_3 = 3'$ and $y_1 = 1.5'$, $y_2 = 4.5'$, $y_1 = 3'$ respectively. Determine the moments about points H and B due to the forces shown.

Reference

Walker, K. M., *Applied Mechanics for Engineering Technologists*, 8th Edition, Prentice Hall, Upper Saddle River, NJ, 2007.

Chapter 4

Two-Dimensional Equilibrium

Objectives: When you complete this chapter you will have:

- Understanding of how to draw free-body diagrams (FBDs) of entire or part mechanisms.
- Understanding of how to apply the three equations of equilibrium: $\sum F_x = 0$; $\sum F_y = 0$; $\sum M = 0$, to FBDs.
- Used SolidWorks to solve equilibrium problems for concurrent, parallel, and non-concurrent systems.

Free-Body Diagrams

FBDs are diagrams of objects in static (not moving) equilibrium (forces acting on the object are balanced against each other so that the object does not move). A FBD shows an object or body with all supports removed and replaced by forces in balance. All forces acting on an object must be shown.

One mistake that students make is to retain the fixtures supporting the body for which a FBD is sought. The guiding principle is that a free-body is indeed 'free'! Students normally enjoy this definition, but end up making mistake when they are drawing FBDs. FBDs must float otherwise they cannot be free-body diagrams.

A FBD of a member is a picture showing how the rest of the world is acting *on* the member, not what the member is doing *to* anything else. This concept is important; visualize yourself as taking the place of a member and consider how all other forces around you are acting on you.

Another pitfall with students is that they do not obey the equations set out for equilibrium. Let us consider the equilibrium equations:

$$\sum F_x = 0; \quad \sum F_y = 0; \quad \sum M = 0.$$

The forces must be summed up with appropriate signs and equated to 0; so also the moment about a joint of the mechanism must be summed up with appropriate signs and equated to 0. If this is not done correctly, it leads to an error. Therefore, it is recommended that the mathematical definitions for equilibrium be enforced when writing the equilibrium equations. Doing things correctly is a good practice that leads to correct solutions.

The other important point to note is that the FBD conventions are extremely useful.

Free-body diagram conventions

There are conventions for replacing supports with equivalent supporting forces as shown in Figure 4.1.

1. *Roller*: The roller cannot exert a horizontal force; therefore, only a force perpendicular to the surface is present (see 1 in Figure 4.1).
2. *Roller*: The only roller force is that perpendicular to the roller surface (see 2 in Figure 4.1).
3. *Smooth Surface*: Zero friction is assumed; therefore, only one force, that perpendicular to the surface, is present (see 3 in Figure 4.1).

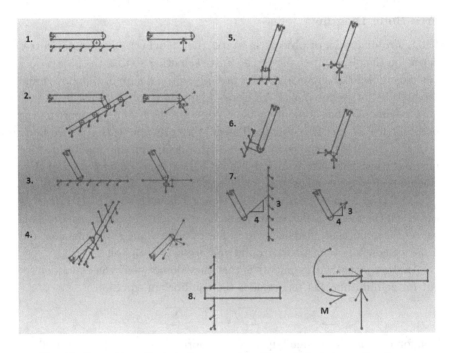

Fig. 4.1 Conventions for replacing supports with equivalent supporting forces.

4. *Slot*: The same principle applies as for a smooth surface. There is only one force present, that perpendicular to the slot (see 4 in Figure 4.1).
5. *Pinned*: Both horizontal and vertical components must be assumed at a pinned connection unless it is on a roller or smooth surface (see 5 in Figure 4.1).
6. The orientation of the support is immaterial; horizontal and vertical components must be assumed (see 6 in Figure 4.1).
7. *Cable*: This is always a single force pulling in the direction of the cable (see 7 in Figure 4.1).
8. *Fixed Support*: The beam is embedded in the wall or support and therefore has three possible reactions: A moment, vertical force, and horizontal force (see 8 in Figure 4.1).

Three Equations of Equilibrium

For complete static equilibrium, three requirements must be met:

1. Horizontal forces balance.
2. Vertical forces balance.
3. Moments balance; clockwise = counter-clockwise (about any point).

Mathematically, these conditions can be stated as follows:

1. $\sum F_x = 0$.
2. $\sum F_y = 0$.
3. $\sum M = 0$.

Two-Force Members

A member that is acted upon by two forces — for example, one at each end — is known as a *two-force member*. A *two-force member* can only be in either *tension* or *compression*.

Pulleys

Pulley systems are used for lifting and lowering loads (see Figure 4.2). Their free-body analysis requires careful consideration.

Problem 4.1

Solve for the forces at A and B of the beam shown in Figure 4.3.

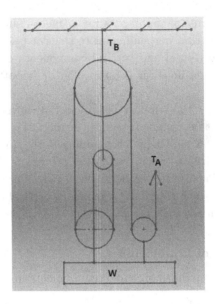

Fig. 4.2 Pulley system.

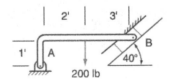

Fig. 4.3 Beam for analysis.

SolidWorks solution

1. Open a **SolidWorks Part** Document.
2. Select the **Front Plane**.
3. Create a **Sketch** for a model (see Figure 4.4).
4. **Exit** Sketch mode.

Create structural members

1. Click **Insert > Weldments > Structural Members** to create Structural members.
2. For **Standard**, select **ansi inch** (see Figure 4.5).

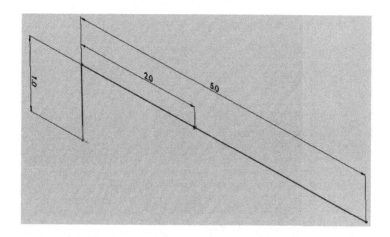

Fig. 4.4 Sketch for creating structural members.

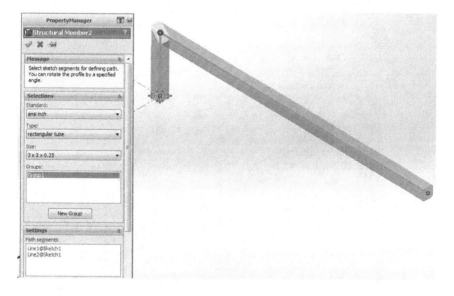

Fig. 4.5 Structural member PropertyManager.

3. For **Type**, select **Rectangular tube** (modified to solid cross-section of
   0.2 × 0.2 × R.05).
4. For **Type**, select **3 × 2 × 0.25**.
5. For **New Group**: Select **Line1, Line2, Line3**.

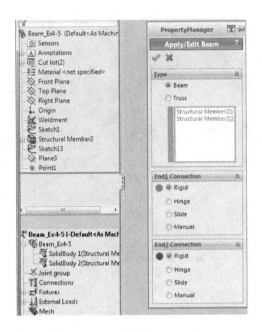

Fig. 4.6 Apply/Edit Beam PropertyManager.

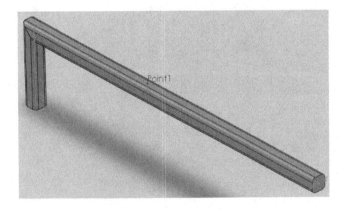

Fig. 4.7 Structural member created.

6. Select **Beam** mode (see Figure 4.6).
7. Click **OK** to complete creating structural member (see Figure 4.7).
8. Create a **Plane** at the right *end of the beam* (see Figure 4.8).
9. Create a Point, **Point1**, 2 feet *from left of the beam* (see Figure 4.9).

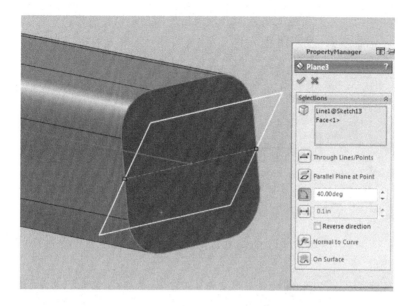

Fig. 4.8 Plane3.

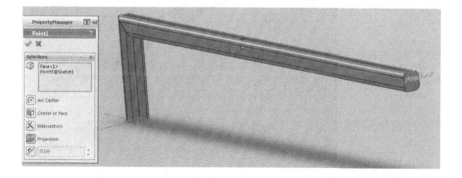

Fig. 4.9 Point1.

Fixtures

1. Right-click on the **Fixtures** tool from SolidWork Simulation.
2. Select **Immovable** condition and the left bottom joint (see Figure 4.10).
3. Select **Use Reference Geometry** condition and the right joint moving along Plane3 (see Figure 4.11).

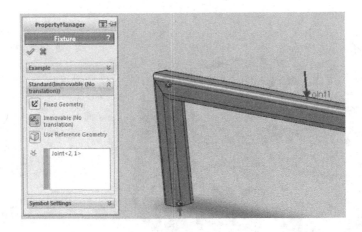

Fig. 4.10 Left-bottom joint end condition.

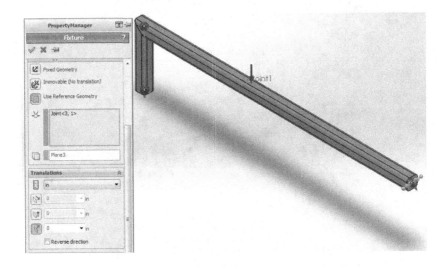

Fig. 4.11 Right joint end condition.

External loads

1. Right-click on the **External Loads** tool from SolidWork Simulation.
2. Apply **20 lb** to Point1 (see Figure 4.12).

Meshing

1. Right-click on the **Mesh** tool from SolidWork Simulation.
2. Click **Create Mesh** (see Figure 4.13 for meshed model).

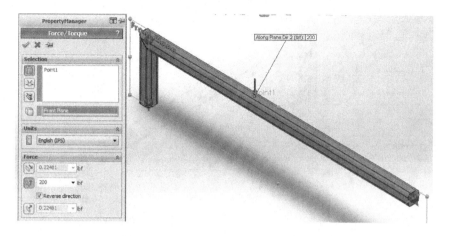

Fig. 4.12 Load applied.

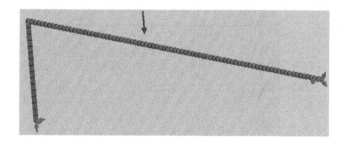

Fig. 4.13 Mesh creation.

Run model

1. Click on the **Run** tool from SolidWork Simulation.
2. Right-click **Results > List Result Force** option.
3. Click the left-bottom joint and right joint (see Reactive Forces in Figure 4.14).

In this simulation, use the file in the Ex4-5A folder. SolidWorks results show the following:

$$A_x = 77\,lb \rightarrow$$
$$A_y = 135\,lb \uparrow$$
$$B = \sqrt{77^2 + 64.6^2} = 101\,lb \ \underline{50°} \uparrow$$

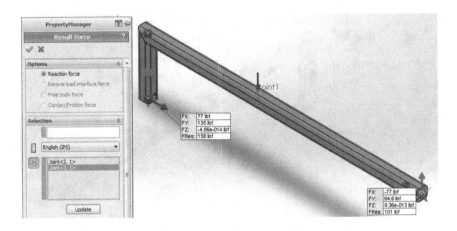

Fig. 4.14 Reactive forces for joints A and B.

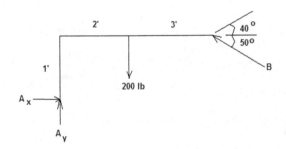

Fig. 4.15 FBD of frame.

Validation

Draw the FBD as shown in Figure 4.15.

$$\sum M_A = 0; \ (B\cos 50)(1) + (B\sin 50)(5) - (200)(2) = 0$$

$$0.643B + 3.83B = 400$$

$$B = 89.4 \, lb \ \underline{50^\circ} \uparrow$$

$$\sum F_x = 0; \ A_x - B\cos 50 = 0$$

$$A_x = 57.5 \, lb \rightarrow$$

$$\sum F_y = 0; \ A_y - 200 + B\sin 50 = 0$$

$$A_y = 132 \, lb \uparrow$$

Table 4.1 Comparative study results.

	SolidWorks [N]	Theory [N]
A_x	77 lb	57.5 lb
A_y	135 lb	132 lb
B	101 lb	89.4 lb

Problem 4.2

A beam has concentrated loads applied as shown in Figure 4.16. Calculate the reactions at A and B.

SolidWorks solution

1. Open a **SolidWorks Part** Document.
2. Select the **Front Plane**.
3. Create a **Sketch1** for a model (see Figure 4.17).
4. **Exit** Sketch mode.

Create structural members

1. Click **Insert > Weldments > Structural Members** to create structural members.
2. For **Standard**, select **ansi inch**.

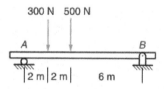

Fig. 4.16 Beam for analysis.

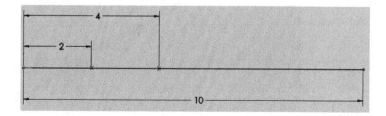

Fig. 4.17 Sketch1 for beam.

3. For **Type**, select **rectangular tube** (modified to solid cross-section of $0.02 \times 0.02 \times$ R.01) (see Figure 4.18).
4. For **Type**, select **3 × 2 × 0.25**.
5. For **New Group**: Select **Line1, Line2, Line3**.
6. Select **Beam** mode.
7. Click **OK** to complete creating structural member.
8. Create a Point, **Point1**, 2 m *from left of the beam* (see Figure 4.19).
9. Create a Point, **Point2**, 4 m *from left of the beam* (see Figure 4.20). (See Point1 and Point2 on model in Figure 4.21.)

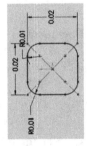

Fig. 4.18 Modified cross-section.

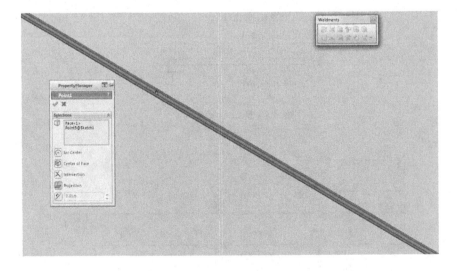

Fig. 4.19 Point1 created.

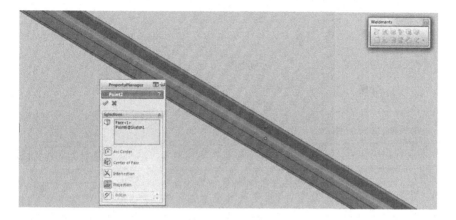

Fig. 4.20 Point2 created.

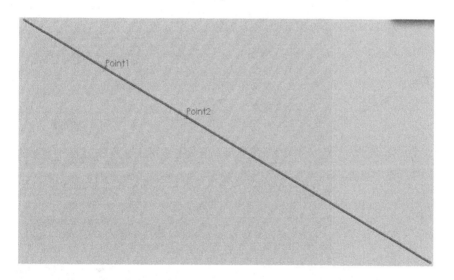

Fig. 4.21 Point1 and Point2 on model.

Fixtures

1. Right-click on the **Fixtures** tool from SolidWork Simulation.
2. Select **Use Reference Geometry** condition and the left joint (see Figure 4.22).
3. Select **Immovable** condition and the right bottom joint (see Figure 4.23).

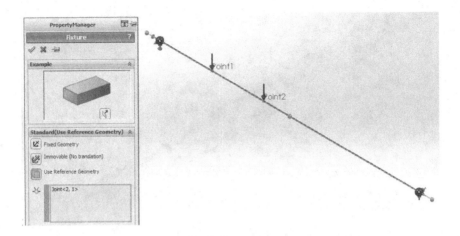

Fig. 4.22 Left joint fixture.

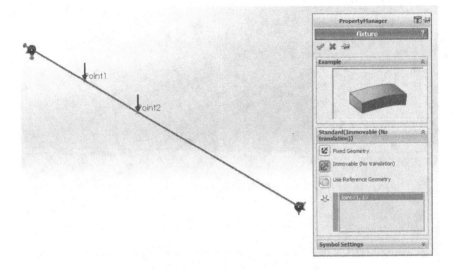

Fig. 4.23 Right joint fixture.

External loads

1. Right-click on the **External Loads** tool from SolidWork Simulation.
2. Apply **300 N** to Point1 (see Figure 4.24).
3. Apply **500 N** to Point2 (see Figure 4.25).

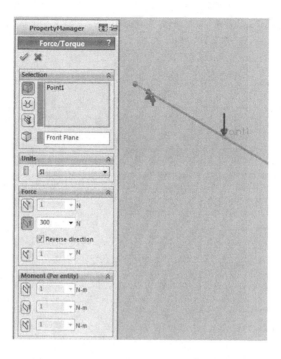

Fig. 4.24 Load on Point1.

Meshing

1. Right-click on the **Mesh** tool from SolidWork Simulation.
2. Click **Create Mesh**.

Run model

1. Click on the **Run** tool from SolidWork Simulation. The default solution is shown in Figure 4.26.
2. Right-click **Results** > **List Result** > **Define Beam Diagrams** option (see Figure 4.27a).
3. From **Display** option, select Shear Force in Dir2 (see Figure 4.27b; see Figure 4.28 for results).
4. Right-click **Results** > **List Beam Forces** option (see Figure 4.29 and Figure 4.30).
5. Right-click **Results** > **Result Force** option.

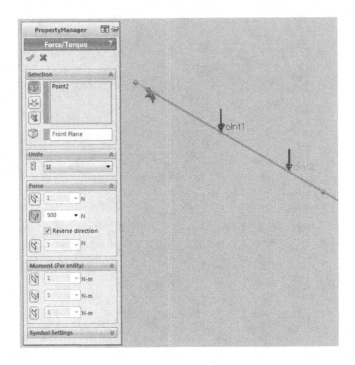

Fig. 4.25 Load on Point2.

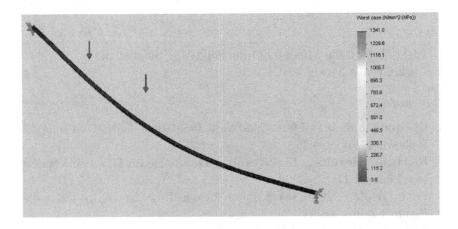

Fig. 4.26 Worst case for model.

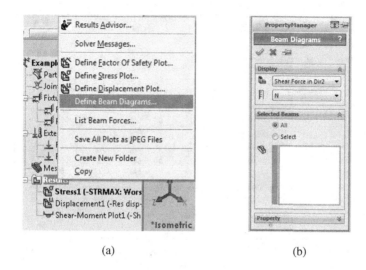

(a)   (b)

Fig. 4.27 Results options.

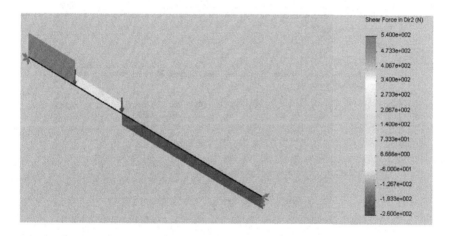

Fig. 4.28 Shear force in Direction2 for beam.

6. Click **Joint < 2, 1 > & Update; Joint < 1, 1 > & Update** option (see Figure 4.31).

SolidWorks results show the following:

$$A_y = 540\,N \uparrow$$
$$B_y = 260\,N \uparrow$$

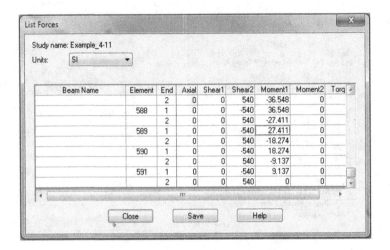

Fig. 4.29 Shear force for joint A (left).

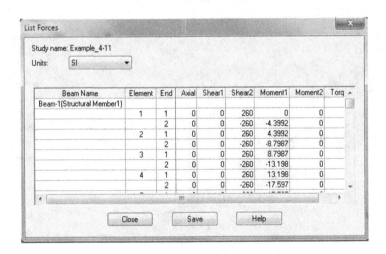

Fig. 4.30 Shear force for joint B (right).

Validation

$$\sum M_A = 0; \ (B_y)(10) - (300)(2) - (500)(4) = 0$$

$$10B_y = 600 + 2000 = 2600$$

$$B_y = 260\,N \uparrow$$

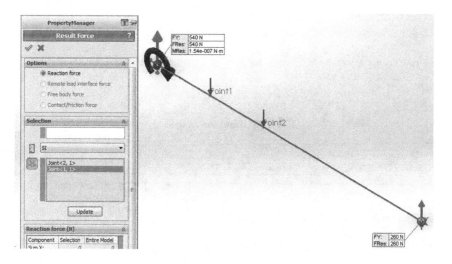

Fig. 4.31 Reactive forces at A and B.

Table 4.2 Comparative study results.

	SolidWorks [N]	Theory [N]
A_y	540	540
B_y	260	260

$$\sum F_y = 0; \ A_y - 300 - 500 + B_y = 0$$
$$A_y = 540 \, N \uparrow$$

Problem 4.3

Solve for reactions A_y and B_y in Figure 4.32.

SolidWorks solution

1. Open a **SolidWorks Part** Document.
2. Select the **Front Plane**.
3. Create a **Sketch1** for a model (see Figure 4.33).

The way the beam model for this problem is created is different from the use of Weldment's method. Notice that the reactive forces are not at

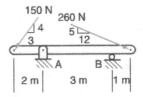

Fig. 4.32 Beam for analysis.

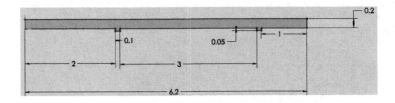

Fig. 4.33 Sketch1 for beam model.

Fig. 4.34 Extrude1 for beam model.

the extreme ends of the beam. Therefore, we create two 'support areas' of finite widths (in this case, a width of 0.1 m as shown in Figure 4.33). We extrude Sketch1 through 0.1 m to form the beam (see Figure 4.34).

The beam is shown in Figure 4.35 with two points (Point1 and Point2) defined.

Point1 is for the left-hand load definition while Point2 is for the right-hand load definition (see Figure 4.36).

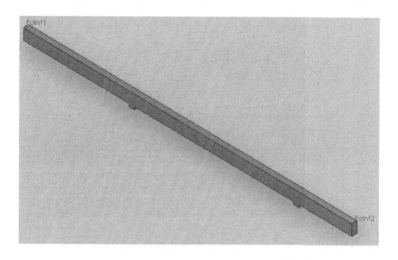

Fig. 4.35 Beam model.

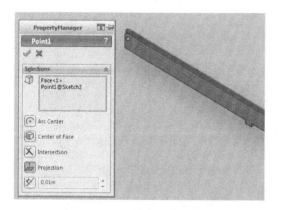 

Fig. 4.36 Point1 and Point2 definitions.

Fixtures

1. Right-click on the **Fixtures** tool from SolidWork Simulation.
2. Select **Fixed Geometry** condition and the left joint (see Figure 4.37).
3. Select **Roller** condition and the right joint (see Figure 4.38).

External loads

1. Right-click on the **External Loads** tool from SolidWork Simulation.

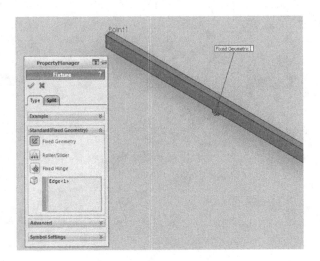

Fig. 4.37 Left joint fixture definition.

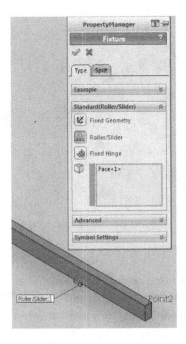

Fig. 4.38 Right joint fixture definition.

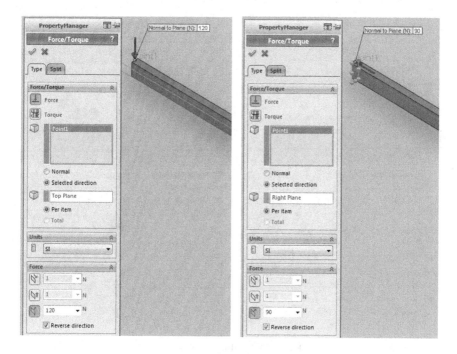

Fig. 4.39 Loading for Point1.

2. Apply **120 N** (downward) and **90 N** (to the left) to Point1 (see Figure 4.39).
3. Apply **100 N** (downward) and **240 N** (to the right) to Point2 (see Figure 4.40).

Meshing

1. Right-click on the **Mesh** tool from SolidWork Simulation.
2. Click **Create Mesh**.

Run model

1. Click on the **Run** tool from SolidWork Simulation.
2. Right-click **Results > Result Force** option (see Figure 4.41).
3. Click **Joint < 1, 1 > & Update** option (see Figure 4.42).
4. Right-click **Results > Result Force** option (see Figure 4.41).
5. Click **Joint < 2, 1 > & Update** option (see Figure 4.43).

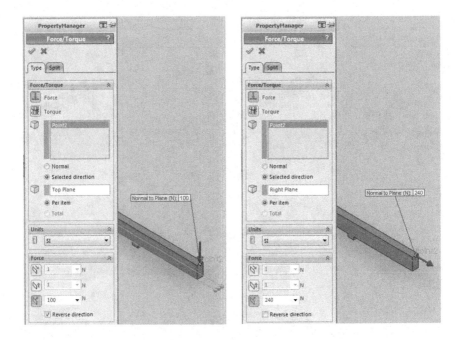

Fig. 4.40 Loading for Point2.

SolidWorks results show the following:

$$A_x = 150\,N \rightarrow$$
$$A_y = 167\,N \uparrow$$
$$B_y = 53.3\,N \uparrow$$

The von Mises stress is shown in Figure 4.44.

Validation

Draw the FBD as shown in Figure 4.45.

$$\sum F_x = 0;\ 240 - 90 - A_x = 0$$
$$A_x = 150\,N \leftarrow$$

$$\sum M_A = 0;\ (120)(2) + (B_y)(10) - (100)(4) = 0$$
$$3B_y = 400 - 240 = 160$$
$$B_y = 53.3\,N \uparrow$$

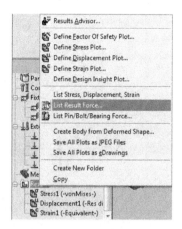

Fig. 4.41 Results option.

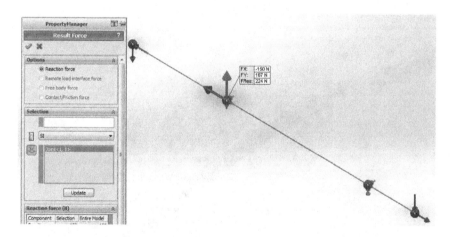

Fig. 4.42 Reactive force at joint A.

$$\sum M_B = 0; \ (120)(5) - (100)(1) - (A_y)(3) = 0$$
$$3A_y = 600 - 100 = 500$$
$$A_y = 167\,N \uparrow$$

Problem 4.4

Solve for reactions A_y and B_y in Figure 4.46.

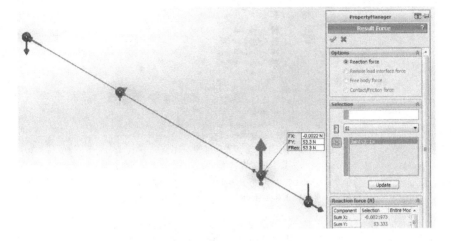

Fig. 4.43 Reactive force at joint B.

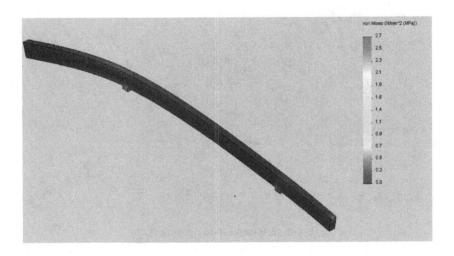

Fig. 4.44 von Mises stress.

SolidWorks solution

1. Open a **SolidWorks Part** Document.
2. Select the **Front Plane**.
3. Create a **Sketch1** for a model (see Figure 4.47).

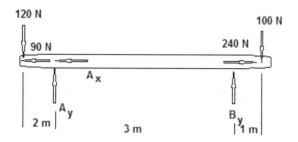

Fig. 4.45 FBD of frame.

Table 4.3 Comparative study results.

	SolidWorks [N]	Theory [N]
A_x	150	150
A_y	167	167
B_y	53.3	53.3

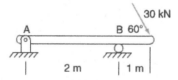

Fig. 4.46 Beam for analysis.

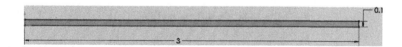

Fig. 4.47 Sketch1 for beam model.

The way the beam model for this problem is created is different from the use of Weldment's method. Notice that the reactive forces are not at the extreme ends of the beam. Sketch1 is created and extruded through 0.05 m to form the beam (see Figure 4.48).

4. Two 'support areas' of finite widths (0.05 m) are created as shown in Figure 4.49.
5. Point1 is created at the right-end of the beam for loading.

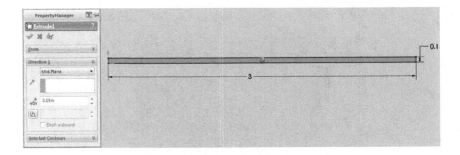

Fig. 4.48 Sketch1 for beam model.

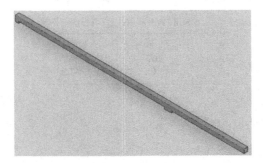

Fig. 4.49 Two 'support areas' of finite widths created.

Fig. 4.50 Creating a mid-line.

6. Create a mid-line to terminate at the end of the beam (see Figure 4.50).
7. Click **Reference Geometry**.
8. Select **Point1@Sketch1** and top-face (see Figure 4.51).

Apply material

1. Right-click **Beam_Ex4-4** and select **Apply/Edit Material** (see Figure 4.52).
2. Apply **Alloy Steel** and Close.

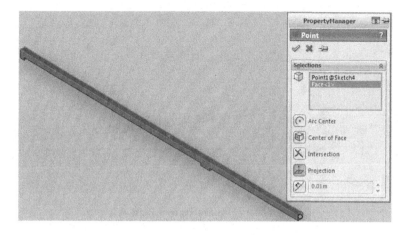

Fig. 4.51 Point1 created.

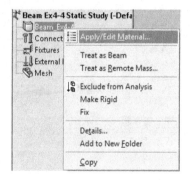

Fig. 4.52 Apply material to beam model.

Fixtures

1. Right-click on the **Fixtures** tool from SolidWork Simulation.
2. Select **Fixed Geometry** condition and the left joint (see Figure 4.53).
3. Select **Roller** condition and the right joint (see Figure 4.54).

External loads

1. Right-click on the **External Loads** tool from SolidWork Simulation.
2. Apply **15000 N** (downward) and **25980 N** (to the right) to Point1 (see Figure 4.55).

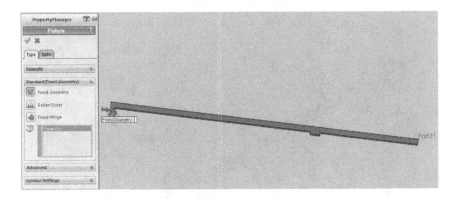

Fig. 4.53 Left joint fixture definition.

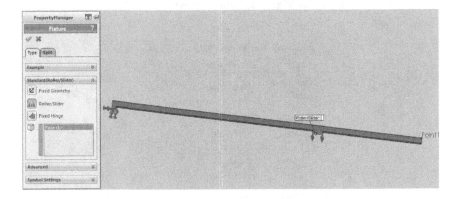

Fig. 4.54 Right joint fixture definition.

Meshing

1. Right-click on the **Mesh** tool from SolidWork Simulation.
2. Click **Create Mesh**.

Run model

1. Click on the **Run** tool from SolidWork Simulation.
2. Right-click **Results > Result Force** option (see Figure 4.56).
3. Click **Face < 1 > & Update; Face < 2 > & Update** option (see Figure 4.57).

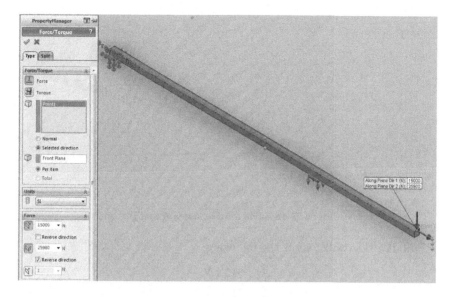

Fig. 4.55 External load application.

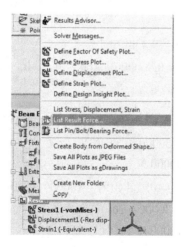

Fig. 4.56 Results option.

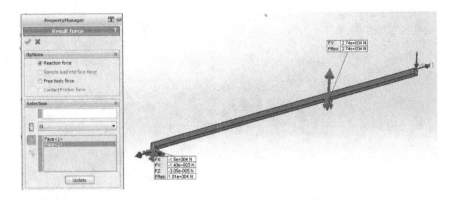

Fig. 4.57 Reactive force at joints A and B.

SolidWorks results show the following:

$$A_x = 15\,kN \rightarrow$$
$$A_y = 1.43\,kN \uparrow$$
$$B_y = 27\,k\,N \uparrow$$

Validation

Draw the FBD as shown in Figure 4.58.

$$\sum F_x = 0; \ 30000\cos 60 - A_x = 0$$
$$A_x = 15000\,N \leftarrow$$

$$\sum M_A = 0; \ (B_y)(2) - (30000\sin 60)(3) = 0$$
$$2B_y = (30000\sin 60)(3)$$
$$B_y = 39000\,N \uparrow$$
$$\sum M_B = 0; \ (A_y)(2) - (30000\sin 60)(1) = 0$$
$$2A_y = (30000\sin 60)(1)$$
$$A_y = 13000\,N \uparrow$$

The von Mises stress is shown in Figure 4.59.

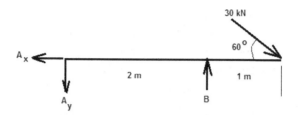

Fig. 4.58 FBD of frame.

Table 4.4 Comparative study results.

	SolidWorks [k N]	Theory [k N]
A_x	15	15
A_y	1.4	13
B_y	27	39

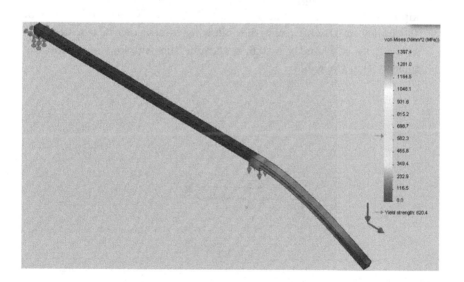

Fig. 4.59 von Mises stress.

Summary

This chapter discusses how SolidWorks is applied to 2D equilibrium problems. The solutions obtained are compared to theoretical values and errors obtained are not significant.

Exercises

P1. Determine the load in each pin-connected member in Figure P1.

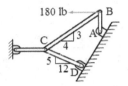

Fig. P1

P2. The boom AC is supported by a cable and pin, with loads $W_1 = 7.87$ kip and $W_2 = 11.24$ kip as shown in Figure P2. Neglecting the size of the pulley at D, and using the following dimensions in feet: $x_1 = 4$; $x_2 = 1$; $x_3 = 5$, and $y = 6.6$, determine the reactions at C and the tension T in the cable.

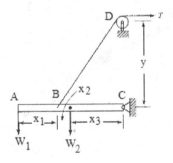

Fig. P2

P3. Determine the reactions at A and B for the beam loaded as shown in Figure P3. Neglect the beam weight.

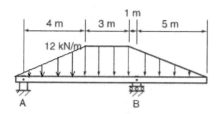

Fig. P3

Reference

Walker, K. M., *Applied Mechanics for Engineering Technologists*, 8th Edition, Prentice Hall, Upper Saddle River, NJ, 2007.

Chapter 5

Structures and Members

Objectives: When you complete this chapter you will have:

- Understanding of the difference between beam and solid elements.
- Understanding of the steps taken in static analysis of beam structures.
- Used *beam elements* in SolidWorks Simulation to carry out static analysis of structural elements such as trusses and machine frames using the concept of weldments.

Truss and Beam Structures

Truss and beam structures are commonly found as structural elements in civil engineering applications. Their applications are common in bridges, where they appear as trusses. Beam structures are also used for design of masts used in oil rigs for supporting drilling operations. They are commonly found in power lines to support cables carrying power, as well as in broadcasting and telecommunication structures as shown in Figure 5.1.

Applied Mechanics textbooks distinguish between trusses, frame, and machines in terms of their configurations and methods of solving these classes of problems. Let us first differentiate them and then review how SolidWorks CAD solves these classes of problems.

Trusses

A truss is a structure made up of straight two-force members that are connected at the joints, with the joints at the ends of the members. They are composed of triangular members connected at joints. Trusses are used to support roofs, bridge, and other structures. The basic structure is triangular or trapezoidal. In a truss, the joints are pin type joints and the members are free to rotate about the pin. As such, a truss cannot transfer moments and members are subjected to only axial forces (tensile and compression).

(a) Phone application   (b) Lighting application

Fig. 5.1 Masts for various applications.

The traditional methods of solving truss problems are:

1. Method of joints.
2. Method of sections.

The *method of joints* consists of a *number of FBDs of adjacent joints*. The first joint selected must have *one or more known external forces* and *only two unknown forces*. A useful approach is to carry out an initial analysis to determine which members are in tension and which are in compression; this approach facilitates finding the solutions in a much easier manner. The unknown forces are determined by using the equilibrium conditions: $\sum F_y = 0$; $\sum F_x = 0$. These newly found forces are used in the FBD of an adjacent joint, to consider subsequent forces in other members. The load in each truss member is found by considering consecutive FBDs throughout the complete truss.

The *method of sections* is used to solve for the forces in a member near the middle of a truss. The method involves cutting a truss into two sections by cutting through the truss where a member force is required. One section is discarded while the FBD of the other is drawn and considered. For both methods, a thumb-rule for designating compressive and tensile forces is that *compressive forces act towards a joint*, while *tensile forces act away from a joint*. The unknown forces are determined by using the equilibrium conditions:$\sum F_y = 0$; $\sum F_x = 0$; $\sum M = 0$.

Frames

On the other hand, members of frames are connected rigidly at joints by means of welding and bolting. Therefore the joints of frames can transfer moments in addition to the axial loads. Frames are structures often composed of pin-connecting multi-force members (i.e. subjected to three or more forces). Unlike trusses, they are not necessarily rigid, for example linked mechanisms.

The traditional method of solving frame problems is *method of members*. Determining the forces acting in each member consists of *drawing FBDs of individual members or of the complete frame.*

Machine members

A machine is also composed of pin-connecting multi-force members, but has moving parts and is usually not considered a rigid structure. Machines are designed to transmit loads rather than support them e.g. a pair of tongs.

Rules for Analysis of Trusses Using Method of Joints

1. Draw the FBD of the entire truss.
2. Determine the reactive forces at the joints that are fixture or moveable.
3. Start analysis with the joint that has at least ONE (1) known external force and NOT MORE than TWO (2) unknown internal (member) forces.
4. Apply equilibrium conditions at the joint ($\sum F_y = 0$; $\sum F_x = 0$).
5. Solve for member force.
6. Progress from joint to joint until all joints have been considered.

Rules for Analysis of Trusses Using Method of Sections

1. Draw the FBD of the entire truss.
2. Determine the reactive forces at the joints that are fixture or moveable.
3. Create Section(s) by cutting NOT more than THREE (3) members.
4. Isolate part of the truss that that has been sectioned.
5. Draw ALL triangles that are cut by the section (dashed lines on the side not required, but full lines on the side required).
6. Seek all joints/vertices where TWO forces meet.

7. Apply equilibrium condition at the vertices where TWO forces meet ($\sum M_{joint} = 0;$).
8. Apply equilibrium conditions to the sectioned truss ($\sum F_y = 0;$ $\sum F_x = 0$).
9. Solve for each member force.

Rules for Analysis of Machine Frames Using Method of Members

1. Draw the FBD of the entire machine frame.
2. Determine the reactive forces at the joints that are fixture or moveable.
3. Draw the FBD of EACH machine member.
4. Apply equilibrium conditions to each machine member ($\sum F_y = 0;$ $\sum F_x = 0; \sum M_{joint} = 0;$).
5. Solve for all reactive forces.

SolidWorks Simulation Procedure

While the traditional approach in Applied Mechanics for solving truss and frame problems is to employ the methods of joints, method of sections, or method of members, SolidWorks CAD system deals with all truss and frame problems in an easy and a unified approach. The key future in the SolidWorks approach is to define each member in a truss or a frame as a *weldment*. In a truss the members are usually defined as a number of triangles that are all connected, but in a frame this may not be the case and care must be taken to model the members appropriately.

The study of weldments is very useful because they are used as a basis for joining structural members in SolidWorks. It is important to know how to deal with beam elements. Loads are applied to joints. Trimming and gussets found in weldments do not necessarily need to be considered when analyzing the structural elements. These simplifications further make beam elements to be easy for analysis.

Beam or truss elements are commonly used in structural members since the number of nodes and elements are greatly reduced (see Figure 5.2). When the length to height ratio (l/h) is greater than or equal to 20:1, beam element is the choice to make. The cross-section of beam elements varies from I-section to hollow rectangular form, depending on the choice. A beam element is very simple compared to the solid element because it is simply a wire-frame as shown in Figure 5.2. Solids require solid elements and shells require shell elements to define meshes.

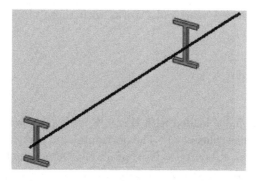

Fig. 5.2 Beam element.

The methodology for beam analysis for trusses and frame structures is briefly given as follows:

1. Open a New Simulation Study.
2. Assign Material Properties (if not already assigned to the model).
3. Carry out Joint Calculation.
4. Assign Fixtures and External Loads.
5. Apply Loads (at joints).
6. Mesh the Model.
7. Run the Analysis.
8. Analyze the Results.

Weldments Toolbar

The **weldments** toolbar provides tools for creating weldment parts. The main tools are:

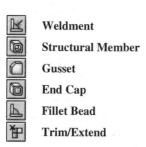

Weldment
Structural Member
Gusset
End Cap
Fillet Bead
Trim/Extend

We will now solve some problems using SolidWorks that belong to the traditional three methods of joints, sections, and members and validate the solutions obtained using SolidWorks with those obtained using these traditional methods.

Method of Joints

Problem 5.1

A truss in Figure 5.3 is loaded with $W_1 = W_2 = W_3 = 4\,kips$ as shown; it is pinned at A and has rollers at F. The horizontal and vertical measurements are $x_1 = x_2 = x_3 = 12'$ and $y = 9'$ respectively. Determine the load in each member of the truss. Material: Steel.

SolidWorks simulation

Create the structural member profile

The first step is to define the paths for which different profiles will be created. The paths are created using 2DSketch.

1. Start a **New SolidWorks Part**.
2. Select the **Front Plane** and be in *Sketch mode*.
3. Sketch the profile for the structure (see Figure 5.4).
4. **Exit** the *Sketch mode*.

Create the weldment structural members

1. Select **Insert > Weldments > Structural Members** in the CommandManager pull-down menu.
2. Click **Structural Member** toolbar.

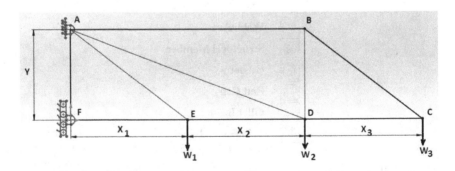

Fig. 5.3 Truss.

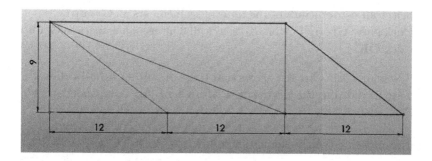

Fig. 5.4 Structural profile.

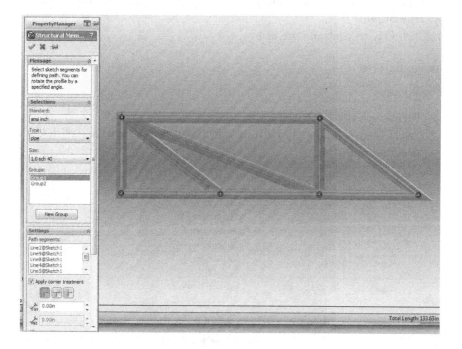

Fig. 5.5 Structural member tool.

3. In the Selection rollout, specify the profile of the structural member by selecting (see Figure 5.5).
4. **Standard**. Select **ansi inch**.
5. **Type**. Select a **Profile Type, pipe**.
6. **Size**. Select a **Profile**, such as **1.0 sch 40**.

7. Select all **6 members** of the outermost members as **Group1**.
8. Click **New Group** button and select all **3** internal members as **Group2**.
9. Click **OK**.

Expanding the **Cut list** shows that there are **nine multi-bodies**; these also are included in the **Structural Member1** if it is expanded as shown in Figure 5.6.

Start a new SolidWorks simulation study

1. Click **Simulation > New Study** (The **SolidWorks Simulation Manager** appears) [Ensure that **Static** Study is selected].
2. Type a name of your choice in the **Name** dialog box, for example *TrussStudyMT*.
3. Click **OK** to continue the Study.

Assign fixtures

Fixed joint

1. Right-click the **Fixtures** folder.
2. Select the *top-left*, apply **Immovable (No translation)** and click **OK** (see Figure 5.7(a)).

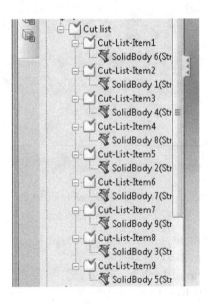

Fig. 5.6 Solid bodies constituting the entire structure.

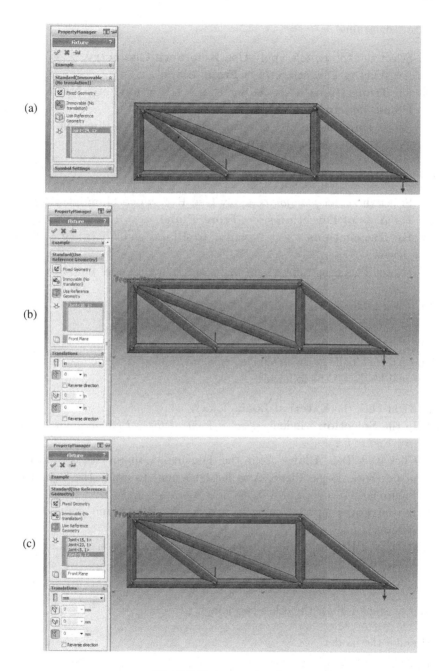

Fig. 5.7 Fixtures assigned.

Roller constraints

3. Select the *bottom-left*, apply **Use Reference Geometry**.
4. For **Axis of Direction** select **Front Plane**.
5. For **Translations** select **Along Plane Direction1** and **Normal to Plane**; **Distances = 0**.
   (Note: Direction1 is in x-direction; Direction2 is in y-direction.)
6. Click **OK** (see Figure 5.7(b)).

Stability constraints

7. Select the *other four joints*, apply **Use Reference Geometry**.
8. For **Axis of Direction** select **Front Plane**.
9. For **Translations** select **Normal to Plane** and **Distance = 0** as shown.
   (Note: This constraint ensures that the structure does not sway; it is stable.)
10. Click **OK** (see Figure 5.7(c)).

Apply external loads

The external loads in the beam elements are applied at the joints. To apply external loads, do the following:

1. Right-click the **External Loads** folder.
2. In the **Selection** window, *select a joint at the bottom* in Joint option (see Figure 5.8).
3. Click the **Top Plane** from the **FeatureManager** as the normal plane (Face, Edge, Plane, Axis for Direction) to apply the loads (see Figure 5.8).
4. In the **Force** option, select **Normal To Plane** and supply 4000 (4 K) psi.
5. Check the **Reverse Direction** to get the right direction of loading (see Figure 5.8).
6. Click **OK**.

Repeat steps 1–6 for the other two joints.

Meshing the model

To mesh the model:
Right-click **Mesh** folder (beam element appears).
Beam elements are wire frames whose cross-sections properties are taken

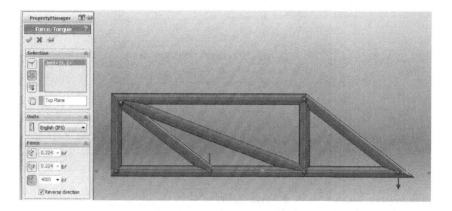

Fig. 5.8 Load application at joints.

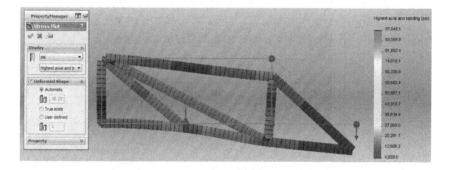

Fig. 5.9 Worst case stress.

from the structural member cross-section geometry defined in SolidWorks during modeling.

Run the analysis

Click the **Run Advisor** in the Simulation CommandManager. The software displays the results and automatically creates two plots in the **Results** folder as follows:

- Worst case stress (see Figure 5.9).
- Resultant displacement.

There are several ways of generating outputs:

1. To *list all beam forces*, **right-click Results** to shows a list of options (see Figure 5.10).

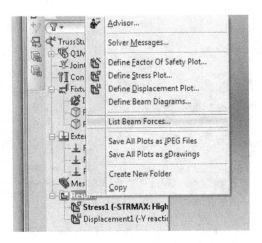

Fig. 5.10 Options to list beam forces and for other plots.

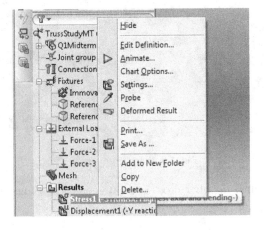

Fig. 5.11 Options for edit definition, change settings, or probe.

2. To *edit definition, change settings, or probe*, **right-click Stress1** under the Results folder to shows a list of options (see Figure 5.11).

3. To show *displacement, reaction force, rotation, reaction moment* for x-, y-, z-directions, and their resultants, **right-click Displacement1** under the Results folder to shows a list of options (see Figure 5.12).

The reactive forces in x-direction, of value 32 kips are shown in Figure 5.13.

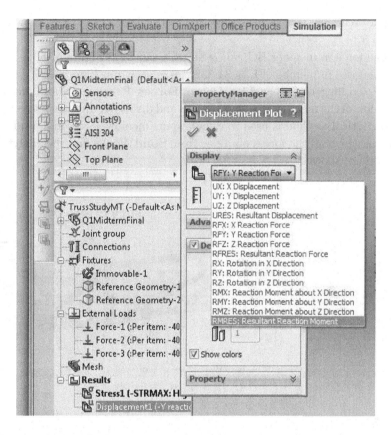

Fig. 5.12 Options for displacement, reaction force, rotation, and reaction moment.

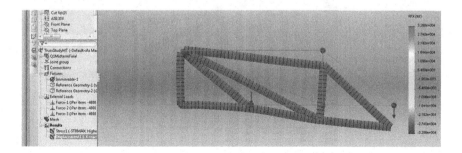

Fig. 5.13 Reactive forces in x-direction.

The reactive forces in y-direction, of value 12 kips are shown in Figure 5.14.

The force in member BC = 6.788 kips is shown in Figure 5.15.

The force in member FE = 31.732 kips is shown in Figure 5.16.

By scrolling down the **List Beam Forces PropertyManager**, the forces for all the members can be accessed and displayed. In the next section, the analytical method is used to calculate the member forces and a table is given in which all the forces obtained from **SolidWorks** and the analytical method are shown.

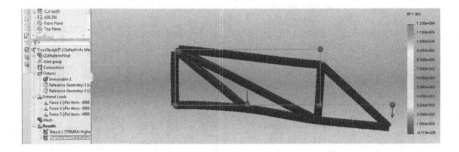

Fig. 5.14 Reactive forces in y-direction.

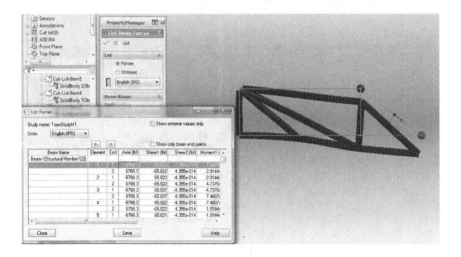

Fig. 5.15 Force in member BC.

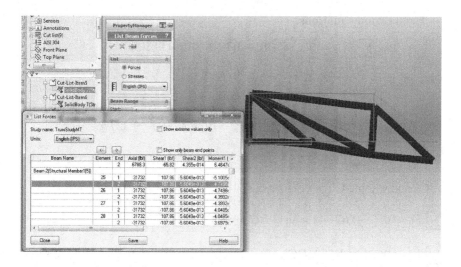

Fig. 5.16 Force in member FE.

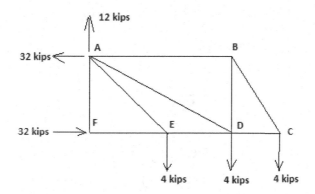

Fig. 5.17 The FBD for the entire structure.

Validation of SolidWorks results

As we already know, it is extremely essential to check that the analyzed results are correct. The best approach is to solve this same problem using the theoretical and cross-check the results from SolidWorks.

The theoretical solution is given as follows: Initial analysis shows that members AF, AB, BC are in tension while members FE, ED, DC are in compression.

An FBD of the entire frame is shown in Figure 5.17.

Equilibrium equations:

$$\sum M_A = 0; \quad F_x(9) - 4(12) - 4(24) - 4(36) = 0$$
$$F_x(9) = 4(12 + 24 + 36) = 4(72)$$
$$\therefore \ F_x = 4(72)/9 = 32 \ kips$$

$$\sum F_x = 0; \quad -A_x + F_x = 0; \quad A_x = F = 32 \ kips$$
$$\sum F_y = 0; \quad A_y - 4 - 4 - 4 = 0; \quad A_y = 12 \ kips$$

1. Consider joint C (see Figure 5.18).

$$\sum F_y = 0; \quad \frac{3}{5}BC - 4 = 0; \quad BC = \frac{5}{3}(4) = 6.67 \ kips \ (T)$$

$$\sum F_x = 0; \quad -\frac{4}{5}BC + DC = 0; \quad DC = \frac{4}{5}BC = \frac{4}{5}(6.67) = 5.336 \ kips \ (C)$$

2. Consider joint B (see Figure 5.19).

$$\sum F_y = 0; \quad BD - \frac{3}{5}BC = 0; \quad BD = \frac{3}{5}(6.67) = 4 \ kips \ (T)$$

$$\sum F_x = 0; \quad -AB + \frac{4}{5}BC = 0; \quad AB = \frac{4}{5}(6.67) = 5.336 \ kips \ (C)$$

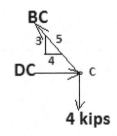

Fig. 5.18 FBD for joint C.

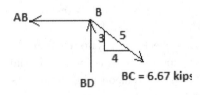

Fig. 5.19 FBD for joint B.

3. Consider joint D (see Figure 5.20).

AD is in tension.

$$\sum F_y = 0; \quad \frac{9}{25.632} AD - 4 - 4 = 0; \quad \frac{9}{25.632} AD = 8;$$

$$AD = \frac{25.632}{9}(8) = 22.784 \; kips \, (T)$$

$$\sum F_x = 0; \quad ED - \frac{24}{25.632} AD - 5.336 = 0;$$

$$ED = \frac{24}{25.632}(22.784) + 5.336 = 26.67 \; kips \, (C)$$

4. Consider joint E (see Figure 5.21).

$$\sum F_y = 0; \quad \frac{3}{5} AE - 4 = 0; \quad AE = \frac{5}{3}(4) = 6.67 \; kips \, (T)$$

$$\sum F_x = 0; \quad FE - \frac{4}{5} AE - ED = 0; \quad FE = \frac{4}{5}(6.67) + 26.67 = 32 \; kips \, (C)$$

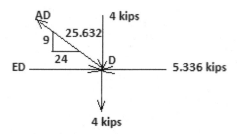

Fig. 5.20 FBD for joint D.

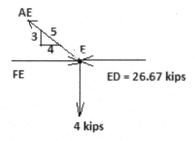

Fig. 5.21 FBD for joint E.

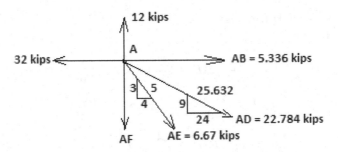

Fig. 5.22 FBD for joint A.

Table 5.1 Results of SolidWorks and the analytical method.

Member	Theoretical method (kips)	SolidWorks (kips)
AB	5.336 (T)	6.205(T)
AD	22.784 (T)	19.297(T)
AE	6.67 (T)	8.921(T)
AF	0	0.107
BD	4 (C)	3.957(C)
BC	6.67 (T)	6.788(T)
FE	32 (C)	31.732(C)
ED	26.67 (C)	24.257(C)
DC	5.336 (C)	5.391(C)

5. Consider joint A (see Figure 5.22).

$$\sum F_y = 0; \quad 12 - AF - \frac{3}{5}(6.67) - \frac{9}{25.632}(22.784) = 0$$
$$AF = 12 - 4 - 8 = 0$$

Table 5.1 shows the results for all the forces obtained from **SolidWorks** and the analytical method.

Method of Sections

Problem 5.2

A pin-connected truss in Figure 5.23 is loaded with $W = 3\,kips$ as shown; it is pinned at D and has rollers at E. The horizontal and vertical measurements are $x_1 = 6'$; $x_2 = 6'$; $x_3 = 4'$ and $y_1 = y_2 = 4'$ respectively. Determine the load in members BC, CE, DE of the truss. Material: Steel.

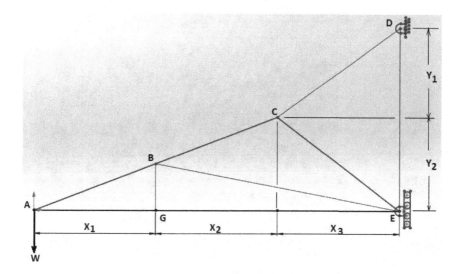

Fig. 5.23 Truss.

SolidWorks simulation

Create the Structural Member Profile

The first step is to define the paths for which different profiles will be created. The paths are created using 2DSketch.

1. Start a **New SolidWorks Part**.
2. Select the **Front Plane** and be in the *Sketch mode*.
3. Sketch the profile for the structure (see Figure 5.24).
4. **Exit** the *Sketch mode*.

Create the weldment structural members

1. Select **Insert > Weldments > Structural Members** in the Com-mandManager pull-down menu.
2. Click **Structural Member** toolbar.
3. In the Selection rollout, specify the profile of the structural member by selecting standard, type, and size:
4. **Standard**. Select **ansi inch**.
5. **Type**. Select a **Profile Type, pipe**.
6. **Size**. Select a **Profile**, such as **1.0 sch 40**.
7. Select all **five members** of the outermost members as **Group1**.

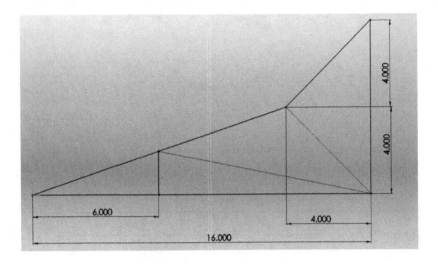

Fig. 5.24 Structural profile.

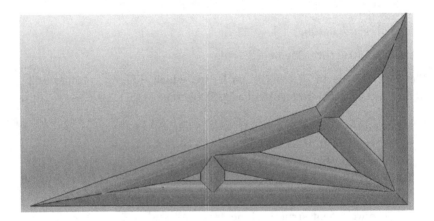

Fig. 5.25 Structural Member tool.

8. Click **New Group** button and select all **three internal members** as **Group2**.
9. Click **OK** (see Figure 5.25 for the defined structural members).

Expanding the **Cut list** shows that there are **eight multi-bodies**; these are also included in the **Structural Member1** if it is expanded as shown in Figure 5.26.

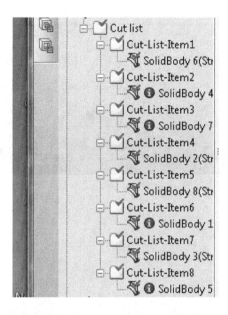

Fig. 5.26 Solid bodies constituting the entire structure.

Start a new SolidWorks simulation study

1. Click **Simulation > New Study** (the **SolidWorks Simulation Manager** appears). (Ensure that **Static** Study is selected.)
2. Type a name of your choice in the **Name** dialog box.
3. Click **OK** to continue the Study.

Assign fixtures

Fixed joint

1. Right-click the **Fixtures** folder.
2. Select the *top-right*, apply **Immovable (No translation)** and click **OK** (see Figure 5.27(a)).

Roller constraints

3. Select the *bottom-right*, apply **Use Reference Geometry**.
4. For **Axis of Direction** select **Front Plane**.
5. For **Translations** select **Along Plane Direction1** and **Normal to Plane; Distances = 0**.

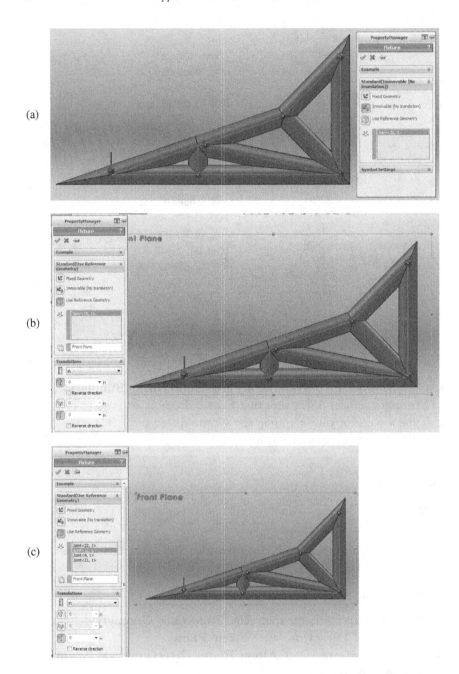

Fig. 5.27 Fixtures assigned.

(Note: Direction1 is in x-direction; Direction2 is in y-direction.)

6. Click **OK** (see Figure 5.27(b)).

Stability constraints

7. Select the *other four joints*, apply **Use Reference Geometry**.
8. For **Axis of Direction** select **Front Plane**.
9. For **Translations** select **Normal to Plane** and **Distance = 0** as shown.
   (Note: This constraint ensures that the structure does not sway; it is stable.)
10. Click **OK** (see Figure 5.27(c)).

Apply external loads

The external loads in the beam elements are applied at the joints. To apply external loads, do the following:

1. Right-click the **External Loads** folder.
2. In the **Selection** window, *select a joint at the bottom* in Joint option (see Figure 5.28).
3. Click the **Top Plane** from the **FeatureManager** as the normal plane (Face, Edge, Plane, Axis for Direction) to apply the loads (see Figure 5.28).
4. In the **Force** option, select **Normal To Plane** and supply 3000 (3 K) psi.

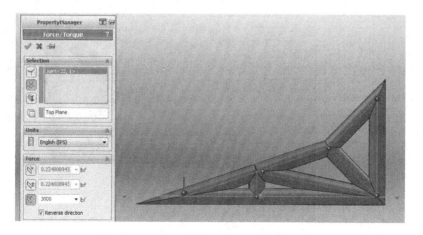

Fig. 5.28 Load application at joints.

5. Check the **Reverse Direction** to get the right direction of loading (see Figure 5.28).
6. Click **OK**.

Meshing the model

To mesh the model:
Right-click **Mesh** folder (beam element appears).
Beam elements are wire frames whose cross-sections properties are taken from the structural member cross-section geometry defined in SolidWorks during modeling.

Run the analysis

Click the **Run Advisor** in the Simulation CommandManager. The software displays the results and automatically creates two plots in the **Results** folder as follows:

- Worst case stress.
- Resultant displacement.

The force in member BC = 9.142 kips is shown in Figure 5.29.

Table 5.2 shows the results for all the forces obtained from **SolidWorks** and analytical method.

Fig. 5.29 Force in member BC.

Table 5.2 Results of **SolidWorks** and analytical method.

Member	Theoretical method (kips)	SolidWorks (kips)
AB	—	6.160 (T)
AE	—	5.789 (C)
BG	—	1.059 (T)
BE	—	1.213 (C)
BC	9.5 (T)	9.142 (T)
CE	4.26 (T)	3.607 (T)
CD	—	8.277 (T)
DE	3.02 (C)	2.603 (C)

Validation of SolidWorks results

Use the method of sections to determine the load in members BC, CE, and DE of the truss.

Initial analysis: members AB, BC, and CD are in tension; AG, and GE are in compression. FBD of entire frame can be seen in Figure 5.29(a).

$$\sum M_D = 0; \quad -E_x(8) + 3(16) = 0$$

$$\therefore E_x = \frac{3(16)}{8} = 6 \ kips$$

$$\sum F_x = 0; \quad D_x - E_x = 0; \ \Rightarrow \ D_x = E_x = 6 \ kips$$

$$\sum F_y = 0; \quad -3 + D_y = 0; \ \Rightarrow \ D_y = 3 \ kips$$

Consider the left-hand side of the section (see Figure 5.29(b)).

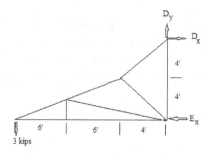

Fig. 5.29(a) Entire frame.

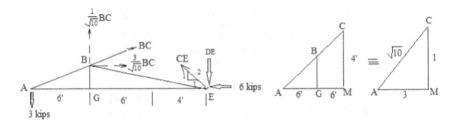

Fig. 5.29(b) Left-hand side of the section and the geometry of the members.

Using similar triangles for the triangles ABG and ACM on the right-hand side of Figure 5.29:

$$\frac{BG}{6} = \frac{4}{12}; \quad \therefore BG = \frac{6(4)}{12} = 2'; \quad AC = \sqrt{3^2 + 1^2} = \sqrt{10}$$

$$\sum M_E = 0; \quad -\frac{3}{\sqrt{10}}BC(2) - \frac{1}{\sqrt{10}}BC(10) + 3(16) = 0$$

$$\therefore BC = 3\sqrt{10} = 9.5 \, kips \, (T)$$

$$\sum F_x = 0; \quad \frac{3}{\sqrt{10}}BC - \frac{1}{\sqrt{2}}CE - 6 = 0$$

$$\therefore CE = 3.01\sqrt{2} = 4.26 \, kips \, (T)$$

$$\sum F_y = 0; \quad -3 + \frac{1}{\sqrt{10}}BC + \frac{1}{\sqrt{2}}CE - DE = 0$$

$$\therefore DE = 3.02 \, kips \, (C)$$

Method of Members

Problem 5.3

Determine the horizontal and vertical components of the pin reactions at A and B on the frame shown in Figure 5.30, in which the loads are $W_1 = 2 \, kips$; $W_2 = 0.5 \, kips$. The horizontal and vertical measurements are $x_1 = 6'$; $x_2 = 2'$; $x_3 = 4'$ and $y_1 = 4'$; $y_2 = 6' = y_3 = 2'$ respectively. Material: Steel.

SolidWorks simulation

Create the Structural Member Profile

The first step is to define the paths for which different profiles will be created. The paths are created using 2DSketch.

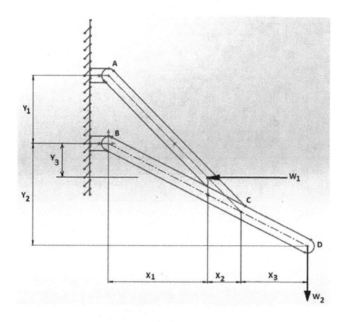

Fig. 5.30 Machine frame.

1. Start a **New SolidWorks Part**.
2. Select the **Front Plane** and be in the *Sketch mode*.
3. Sketch the profile for the structure (see Figure 5.31).
4. **Exit** the *Sketch mode*.

Create the weldment structural members

1. Select **Insert > Weldments > Structural Members** in the CommandManager pull-down menu.
2. Click **Structural Member** toolbar.
3. In the Selection rollout, specify the profile of the structural member by selecting standard, type, and size:
4. **Standard**. Select **ansi inch**.
5. **Type**. Select a **Profile Type, rectangular tube**.
6. **Size**. Select a **Profile**, such as **3 × 2 × 0.25**.
7. Select **lower member** as **Group1**.
8. Click **New Group** button and select **upper member** as **Group2**.
9. Click **OK** (see Figure 5.32 for the defined structural member).

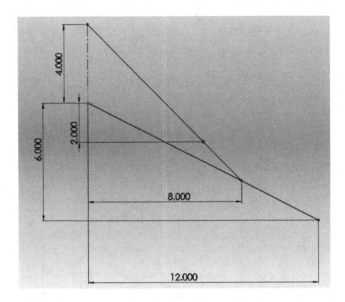

Fig. 5.31 Structural profile.

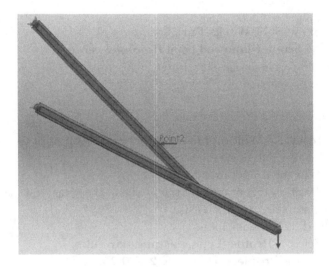

Fig. 5.32 Structural Member tool.

Fig. 5.33 Solid bodies constituting the entire structure.

Expanding the **Cut list** shows that there are **two multi-bodies**; these are also included in the **Structural Member1** if it is expanded as shown in Figure 5.33.

Start a new SolidWorks simulation study

1. Click **Simulation > New Study** (The **SolidWorks Simulation Manager** appears). (Ensure that **Static** Study is selected.)
2. Type a name of your choice in the **Name** dialog box.
3. Click **OK** to continue the Study.

Assign fixtures

Fixed joints

1. Right-click the **Fixtures** folder.
2. Select the *top-left*, apply **Immovable (No translation)** and click **OK** (see Figure 5.34(a)).
3. Select the *bottom-left*, apply **Immovable (No translation)** and click **OK** (see Figure 5.34(b)).

Apply external loads

The external loads in the beam elements are applied at the joints. To apply external loads, do the following:

1. Right-click the **External Loads** folder.
2. In the **Selection** window, select Point2 on the upper member in Vertices, Points option (see Figure 5.35).
3. Click the **Right Plane** from the **FeatureManager** as the normal plane (Face, Edge, Plane, Axis for Direction) to apply the loads (see Figure 5.35).
4. In the **Force** option, select **Normal To Plane** and supply 2000 (2 K) psi.

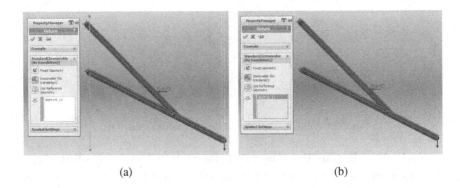

(a)   (b)

Fig. 5.34 Fixtures assigned.

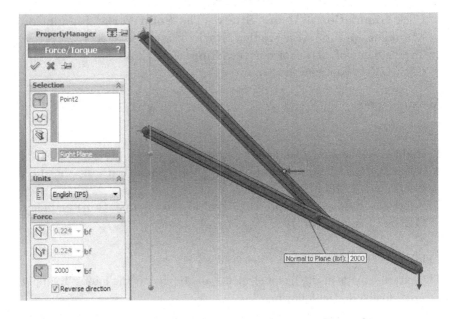

Fig. 5.35 Load application at joints: Horizontal.

5. Check the **Reverse Direction** to get the right direction of loading (see Figure 5.35).

6. Click **OK**.

Apply a second load as follows:

7. Right-click the **External Loads** folder.

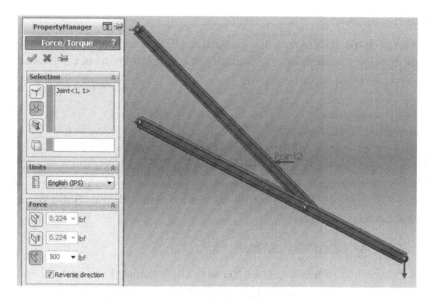

Fig. 5.36 Load application at joints: Vertical.

8. In the **Selection** window, *select a joint at the bottom* in Joint option (see Figure 5.36).
9. Click the **Top Plane** from the **FeatureManager** as the normal plane (Face, Edge, Plane, Axis for Direction) to apply the loads (see Figure 5.36).
10. In the **Force** option, select **Normal To Plane** and supply 500 (0.5 K) psi.
11. Check the **Reverse Direction** to get the right direction of loading (see Figure 5.36).
12. Click **OK**.

Meshing the model

To mesh the model:

Right-click **Mesh** folder (beam element appears).

Beam elements are wire frames whose cross-sections properties are taken from the structural member cross-section geometry defined in SolidWorks during modeling.

Run the analysis

Click the **Run Advisor** in the Simulation CommandManager. The software displays the results and automatically creates two plots in the **Results** folder as follows:

- Worst case stress.
- Resultant displacement.

The reactive forces in the x-direction at A and B are shown in Figure 5.37. The reactive forces in the y-direction at A and B are shown in Figure 5.38. Table 5.3 shows the results for all the forces obtained from **SolidWorks** and analytical method.

Validation of SolidWorks results

Use the method of members to determine the horizontal and vertical components of the pin reactions at A and B on the frame.

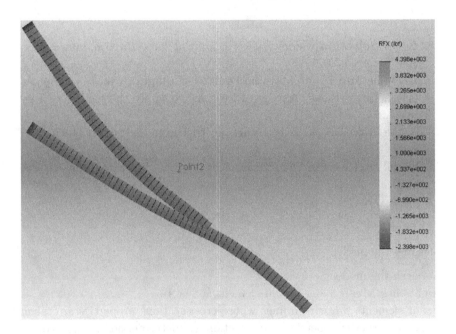

Fig. 5.37 Reactive forces in the x-direction at A and B.

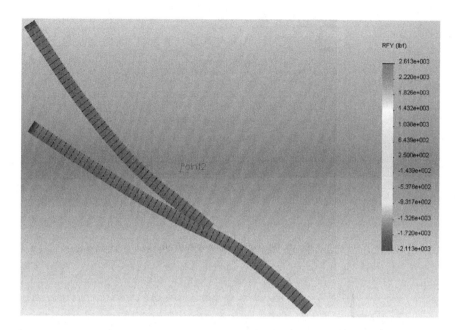

Fig. 5.38 Reactive forces in the y-direction at A and B.

Table 5.3 Results of **SolidWorks** and analytical method.

Member	Theoretical method (kips)	SolidWorks (kips)
A_x	2.5	2.4
A_y	3	2.6
B_x	4.5	4.4
B_y	2.5	2.1

FBD of entire frame (see Figure 5.38(a)):

$$\sum M_B = 0; \quad A_x(4) - 2(2) - 0.5(12) = 0$$
$$\therefore A_x = 2.5 \, kips \; \rightarrow$$

$$\sum F_x = 0; \quad -A_x + B_x - 2 = 0$$
$$\therefore B_x = 4.5 \, kips \; \rightarrow$$

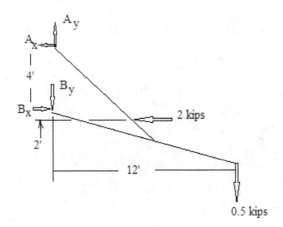

Fig. 5.38(a) FBD of entire frame.

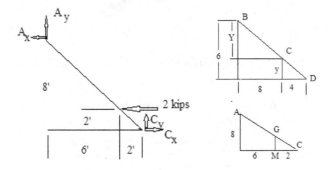

Fig. 5.38(b) FBD of frame AC and the geometry of the members.

FBD of frame AC (see Figure 5.38(b)):

From the triangles on the right-side of Figure 5.38(b), based on the given structure:

$$\frac{y}{4} = \frac{6}{12}; \quad \therefore \; y = 2'; \quad Y + y = 6; \quad \therefore \; Y = 6 - 2 = 4'$$

$$\frac{GM}{2} = \frac{8}{8}; \quad \therefore \; GM = 2'$$

$$\sum M_c = 0; \quad 2(2) + A_x(8) - A_y(8) = 0$$

$$A_y(8) = 2(2) + 2.5(8) = 24$$

$$A_y = 3 \; kips \; \uparrow$$

From FBD of the entire frame:

$$\sum F_y = 0; \quad A_y - B_y - 0.5 = 0$$
$$\therefore \ B_y = A_y - 0.5 = 2.5 \, kips \ \downarrow$$

Problem 5.4

Determine the horizontal and vertical components of the pin reaction at D of the system shown in Figure 5.39, in which the loads are $W_1 = 0.8 \, kips$; $W_2 = 0.5 \, kips$. The horizontal and vertical measurements are $x_1 = 3'$; $x_2 = 5'$; $x_3 = 3'$; $x_4 = 6'$ and $y_1 = 4'$; $y_2 = 8'$ respectively.

SolidWorks simulation

Create the structural member profile

1. Start a **New SolidWorks Part**.
2. Select the **Front Plane** and be in the *Sketch mode*.
3. Sketch the profile for the structure (see Figure 5.40).
4. **Exit** the *Sketch mode*.

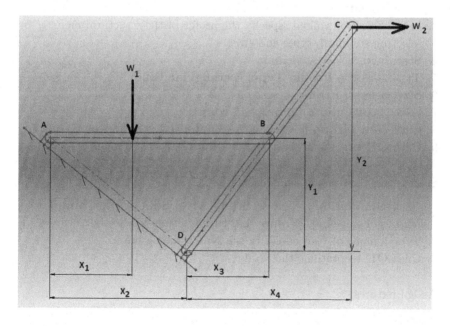

Fig. 5.39 Machine frame.

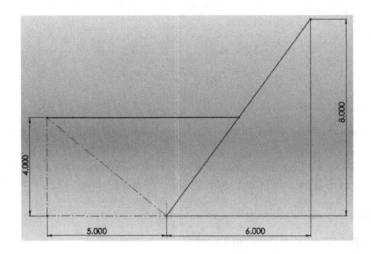

Fig. 5.40 Structural profile.

Create the weldment structural members

1. Select **Insert > Weldments > Structural Members** in the Com-mandManager pull-down menu.
2. Click **Structural Member** toolbar.
3. In the Selection rollout, specify the profile of the structural member by selecting standard, type, and size:
4. **Standard.** Select **ansi inch**.
5. **Type.** Select a **Profile Type, rectangular tube**.
6. **Size.** Select a **Profile**, such as **3 × 2 × 0.25**.
7. Select **horizontal member** as **Group1**.
8. Click **New Group** button and select **slanted member** as **Group2**.
9. Click **OK** (see Figure 5.41 for the defined structural member).

Start a new SolidWorks simulation study

1. Click **Simulation > New Study** (The **SolidWorks Simulation Manager** appears). (Ensure that **Static** Study is selected.)
2. Type a name of your choice in the **Name** dialog box.
3. Click **OK** to continue the Study.

Assign fixtures

Fixed joints

4. Right-click the **Fixtures** folder.

5. Select the *top-left*, apply **Fixed Geometry** and click **OK** (see Figure 5.42).

6. Select the *bottom-left*, apply **Fixed Geometry** and click **OK** (see Figure 5.42).

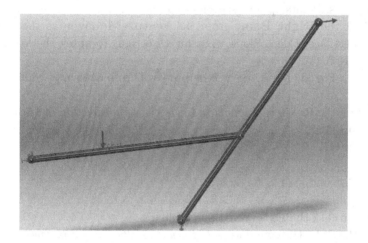

Fig. 5.41 Structural Member tool.

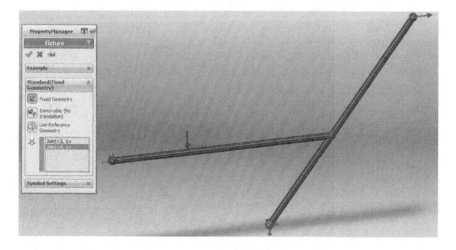

Fig. 5.42 Fixtures assigned.

Apply external loads

The external loads in the beam elements are applied at the joints. To apply external loads, do the following:

1. Right-click the **External Loads** folder.
2. In the **Selection** window, select Point1 on the horizontal member in Vertices, Points option (see Figure 5.43).
3. Click the **Right Plane** from the **FeatureManager** as the normal plane (Face, Edge, Plane, Axis for Direction) to apply the loads (see Figure 5.43).
4. In the **Force** option, select **Normal To Plane** and supply 800 (0.8 K) psi.
5. Check the **Reverse Direction** to get the right direction of loading (see Figure 5.43).
6. Click **OK**.

Apply a second load as follows:

7. Right-click the **External Loads** folder.
8. In the **Selection** window, *select a joint at the slanted member* in Joint option (see Figure 5.44).

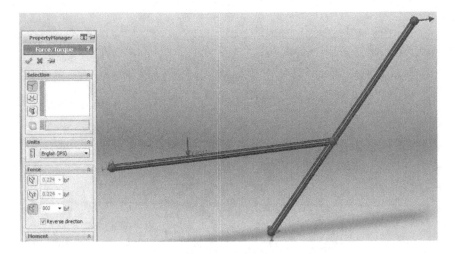

Fig. 5.43 Load application at joints: Vertical.

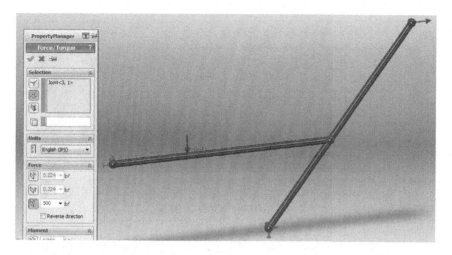

Fig. 5.44 Load application at joints: Horizontal.

9. Click the **Top Plane** from the **FeatureManager** as the normal plane (Face, Edge, Plane, Axis for Direction) to apply the loads (see Figure 5.44).

10. In the **Force** option, select **Normal To Plane** and supply 500 (0.5 K) psi.

11. Check the **Reverse Direction** to get the right direction of loading (see Figure 5.44).

12. Click **OK**.

Meshing the model

To mesh the model:

Right-click **Mesh** folder (beam element appears).

Beam elements are wire frames whose cross-sections properties are taken from the structural member cross-section geometry defined in SolidWorks during modeling.

Run the analysis

Click the **Run Advisor** in the Simulation CommandManager. The software displays the results and automatically creates two plots in the **Results** folder as follows:

- Worst case stress.
- Resultant displacement.

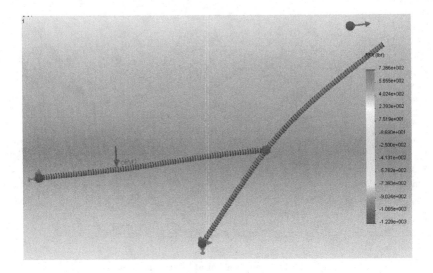

Fig. 5.45 Reactive force in the x-direction at D.

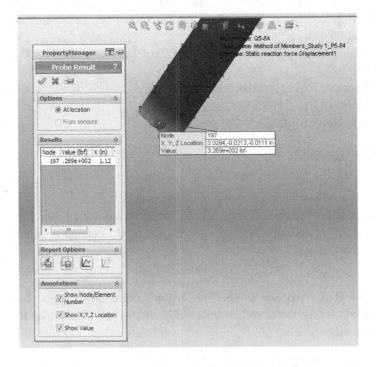

Fig. 5.46 Reactive force in the y-direction at D.

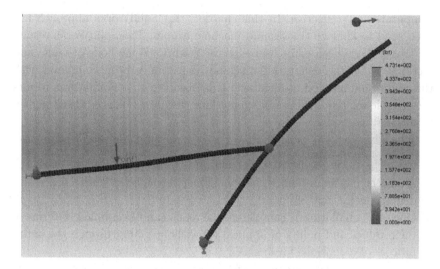

Fig. 5.47 Reactive force in the y-direction at DA.

Table 5.4 Results of SolidWorks and analytical method.

Member	Theoretical method (lb)	SolidWorks (lb)
D_x	725	727
D_y	300	327
A_x	—	1230
A_y	—	473

The reactive force in the x-direction at D is shown in Figure 5.45; the value is 727 lb.

The reactive force obtained using the Probe tool in the y-direction at D is shown in Figure 5.46; the value is 327 lb.

The reactive force in the y-direction at A is shown in Figure 5.47; the value is 473 lb. The reactive force in the x-direction at A is 1230 lb.

Table 5.4 shows the results for all the forces obtained from **SolidWorks** and analytical method.

Summary

Beam elements have been used to analyze weldments which define structural members. It is observed that beam elements are quite simple and computationally effective. In this chapter, this SolidWorks approach has been used to analyze trusses and machine frames. The SolidWorks results for trusses have been compared with the traditional methods of joints and sections, while the results for machine frames have been compared with the traditional method of members. It is observed that SolidWorks results are reliable for practical applications since they are validated using known traditional methods.

The advantages of using SolidWorks for structural analysis are immense:

1. Once the CAD model is completed and the simulation completed, the geometry can be modified without going through the entire process of definitions for analysis.
2. Once the simulation is completed, the loads can be modified in terms of their values and points of application without going through the entire process of definitions for analysis.
3. In other words, maintenance of SolidWorks simulation features is very effective and efficient and is very cost-effective in the long-run.

Exercises

P1. A truss in Figure P1 is loaded with $W_1 = W_2 = W_3 = 3\,kips$ as shown; it is pinned at A and has rollers at F. The horizontal and vertical measurements are $x_1 = x_2 = x_3 = 4'$ and $y = 3'$ respectively. Determine the load in each member of the truss. Material: Steel.

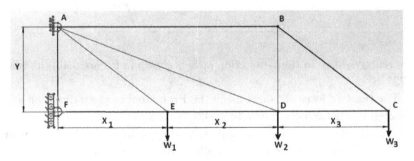

Fig. P1 Truss.

P2. A pin-connected truss in Figure P2 is loaded with $W = 4\,kips$ as shown; it is pinned at D and has rollers at E. The horizontal and vertical measurements are $x_1 = 3'$; $x_2 = 3'$; $x_3 = 2'$ and $y_1 = y_2 = 2'$ respectively. Determine the load in members BC, CE, DE of the truss. Material: Steel.

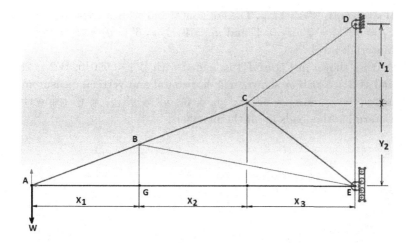

Fig. P2 Truss.

P3. Determine the horizontal and vertical components of the pin reactions at A and B on the frame shown in Figure P3, in which the loads are

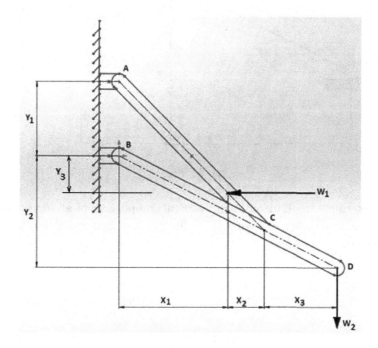

Fig. P3 Machine frame.

$W_1 = 3\,kips$; $W_2 = 1\,kip$. The horizontal and vertical measurements are $x_1 = 6'$; $x_2 = 2'$; $x_3 = 4'$ and $y_1 = 4'$; $y_2 = 6' = y_3 = 2'$ respectively. Material: Steel.

P4. A truss shown in Figure P4 is loaded with $W1 = 950\,lb$, $W2 = 280\,lb$ and $W3 = 500\,lb$ as shown. The horizontal and vertical measurements are $x_1 = x_2 = x_3 = 6'$ and $y_1 = 3'$, $y_2 = 9'$, $y_3 = 6'$ respectively. Determine the loads in all the members.

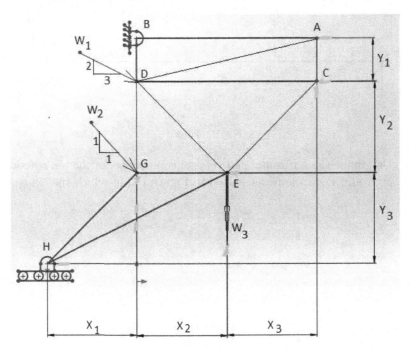

Fig. P4 Truss.

P5. Determine the horizontal and vertical components of all forces acting on each member of Figure P5. (Clearly show how you determine the length CD.)

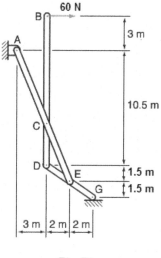

Fig. P5

P6. Determine the horizontal and vertical components of all forces acting on each member of Figure P6.

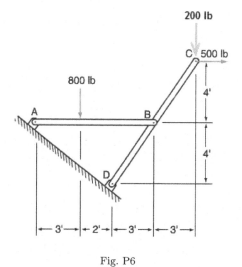

Fig. P6

Reference

Walker, K. M., *Applied Mechanics for Engineering Technolgy*, Prentice Hall, 8th Edition, Prentice Hall, 2007.

Chapter 6

Three-Dimensional Equilibrium

Objectives: When you complete this chapter you will have:

- Understanding of how SolidWorks can be applied to analyze concurrent three-dimensional systems.
- Used SolidWorks to analyze concurrent 3D systems.

Equilibrium in 3D

Forces in space are non-coplanar, and may be one of the following three:

1. Parallel.
2. Non-concurrent.
3. Concurrent.

For forces in 3D, the equilibrium conditions are:

$$\sum F_x = 0; \quad \sum M_x = 0$$
$$\sum F_y = 0; \quad \sum M_y = 0$$
$$\sum F_z = 0; \quad \sum M_z = 0$$

Problem 6.1

Determine the loads in members AD, BD, and CD shown in the system of Figure 6.1. Joints A, B, and C are ball and socket joints.

SolidWorks solution

Analyzing 3D structures analytically is quite challenging. However, using SolidWorks makes the task much easier. The first step is to create the 3D model using the 3D Sketching tool.

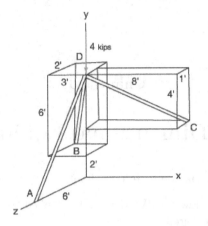

Fig. 6.1 3D force system.

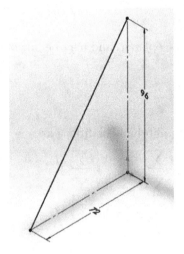

Fig. 6.2 3DSketch1.

1. Open a **New SolidWorks Part** Document.
2. Be in **Sketch** mode.
3. Create the **3DSketch1** shown in Figure 6.2.
4. **Exit Sketch** mode.
5. Create the **3DSketch2** shown in Figure 6.3.
6. **Exit Sketch** mode.
7. Create the **3DSketch3** shown in Figure 6.4.
8. **Exit Sketch** mode.

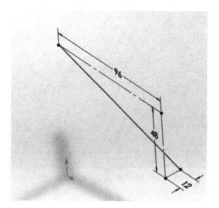

Fig. 6.3 3DSketch2.

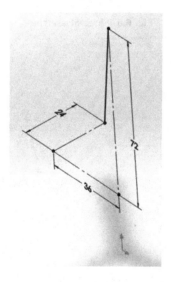

Fig. 6.4 3DSketch3.

The three 3DSketches are shown in Figure 6.5.

Create structural elements

1. Click **Insert > Weldments > Structural Member**.
   (The Structural Member PropertyManager automatically appears; see Figure 6.6)

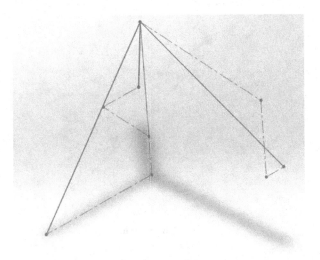

Fig. 6.5 Three 3Dsketches.

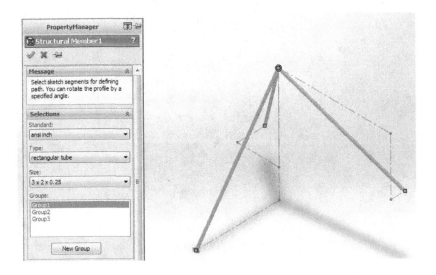

Fig. 6.6 Structural Member PropertyManager.

For Selections, choose the following:

2. **Standard: ansi inch.**
3. **Type: rectangular tube.**
4. **Size: 3 × 2 × 0.25** (this is modified; see structural member).

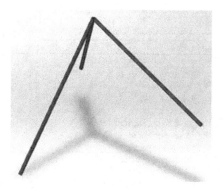

Fig. 6.7 3D model.

5. **Groups**: Select **3DSketch1** as **Group1**; **3DSketch2** as **Group2**; **3DSketch3** as **Group3**.
6. Click **OK** (see Figure 6.7 for the model required for analysis).

Analysis

1. Click **Add-Ins**.
2. Select **SolidWorks Simulation** (see Figure 6.8).
3. Click **Simulation > Study** (1) (see Figure 6.9). (Study PropertyManager is automatically displayed.)
4. Give name as **3D Equilibrium Study** (see Figure 6.10).
5. Click **OK**.

Apply material, fixtures, external loads, and mesh

Material, fixtures, external loads, and mesh are now applied to the model using the CommandManager interface or using the SolidWorks Simulation Manager shown in Figure 6.11.

Apply material

1. Select **Apply Material** [1] (see Figure 6.11).
2. Select **English (IPS)**; **Apply** and **Close** (see Figure 6.12).

Apply fixtures

1. Select **Fixture Advisor** [2] (see Figure 6.11).
2. Select **Immovable (No translation)** and apply the *three lower joints* [**Joint<4, 1>**, **Joint<7, 1>**, **Joint<2, 1>**] (see Figure 6.13).
3. Click **OK**.

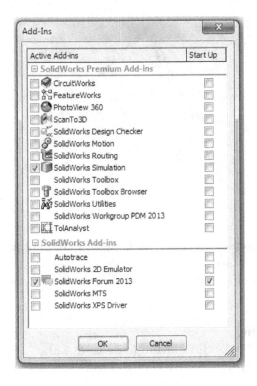

Fig. 6.8 Add-Ins interface.

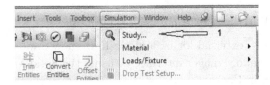

Fig. 6.9 Starting the simulation study.

Apply external loads

1. Select **External loads** [3] (see Figure 6.11).
   (The Force/Torque PropertyManager is automatically displayed; see Figure 6.14)
2. Select *topmost node* [**Joint<5, 1>**].
3. Select **Top Plane** as reference plane.
4. Select for **Units: English (IPS)**.

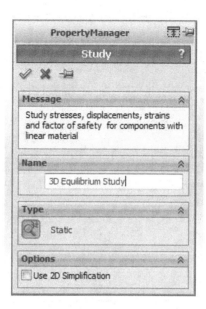

Fig. 6.10 Naming the study.

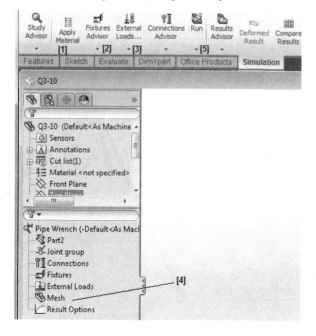

Fig. 6.11 Steps in the Simulation process.

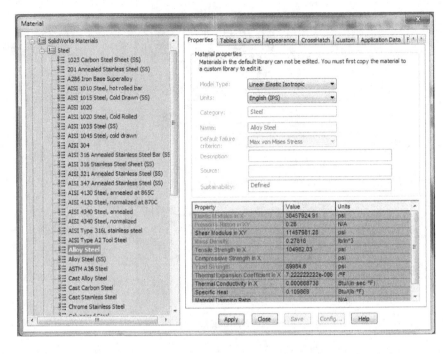

Fig. 6.12 Material selection.

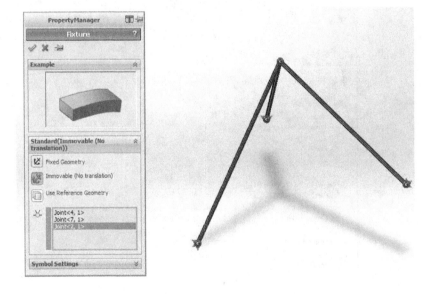

Fig. 6.13 Fixed joint (upper joint).

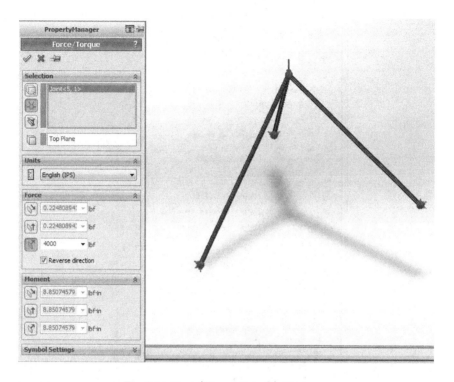

Fig. 6.14 Force/Torque PropertyManager.

5. Apply *normal force* relative to the *Top Plane*, with value = **4000 lb**.
6. Click **OK** to apply.

Apply mesh and run

1. Right-click **Mesh** [4] (see Figure 6.11).
2. Select **Run** [5] (see Figure 6.11).

Viewing results

1. Right-click **Results**.
2. Select **List Beam Forces**... [1] (see Figure 6.15).

The **List Bean Forces PropertyManager** appears automatically as shown in Figure 6.16.

Click **OK** to show the **List Force PropertyManager** (see Figure 6.17 to Figure 6.19).

Fig. 6.15 Selecting result options.

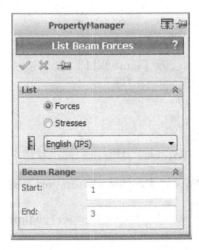

Fig. 6.16 List Bean Forces PropertyManager.

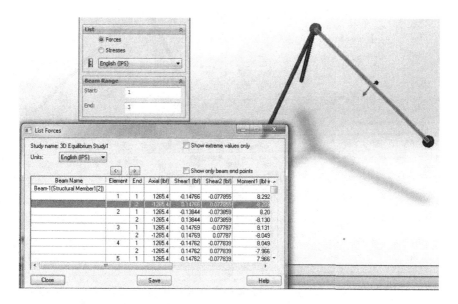

Fig. 6.17 Load on member CD.

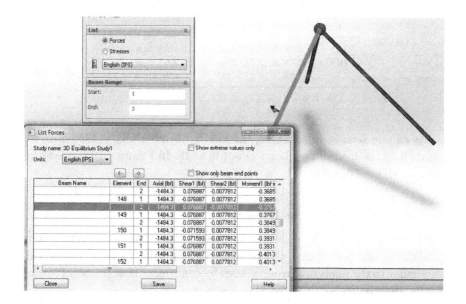

Fig. 6.18 Load on member AD.

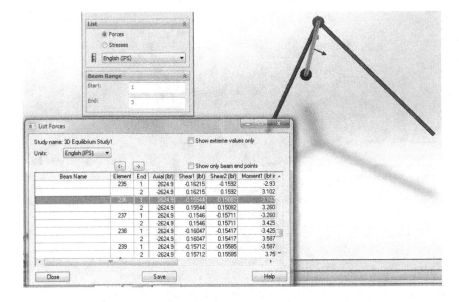

Fig. 6.19 Load on member BD.

Table 6.1 SolidWorks and analytical results.

Member	SolidWorks results	Analytical results
AD	1484.3 (1.48 kips)	1.48 kips
BD	2624.9 (2.62 kips)	2.63 kips
CD	1265.4 (1.27 kips)	1.27 kipjs

As can be seen in Figure 6.17, the load in member CD = 1265.4 lb. In Figure 6.18 the load in member AD = 1484.3 lb. In Figure 6.19 the load in member BD = 2624.9 lb.

Validation of SolidWorks Results

Validation of SolidWorks results is extremely important to determine how good the results are. This step should not be skipped. The results obtained are similar to the analytical results, thereby confirming that the SolidWorks results are reliable.

Summary

This chapter has presented SolidWorks as a useful analysis tool for 3D structures which are ordinarily challenging to analyze using analytical method. The results are found to be very reliable and hence SolidWorks is an alternative analysis tool for these mechanical elements commonly found in practice.

Exercises

P1. Determine the load in each member of the frame shown in Figure P1.

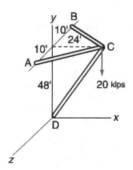

Fig. P1

P2. Determine the load in each member of the frame shown in Figure P2.

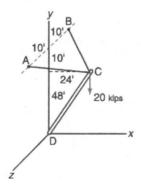

Fig. P2

P3. Determine the load in each member of the frame shown in Figure P3.

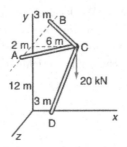

Fig. P3

P4. Determine the load in each member of the frame shown in Figure P4.

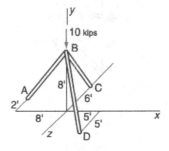

Fig. P4

Reference

Walker, K. M., *Applied Mechanics for Engineering Technologists*, 8th Edition, Prentice Hall, Upper Saddle River, NJ, 2007.

Chapter 7

Friction

Objectives: When you complete this chapter you will have:

- Understanding of applying friction laws for dry surfaces.
- Used SolidWorks to the problem of friction.

Introduction

Friction occurs whenever two bodies are in contact with each other and are either moving with respect to each other or due to forces acting on the bodies causing them to move. When studying friction one has to distinguish between at least two major phenomena:

- **Dry friction** occurs when the surfaces in contact with each other are free of any lubricants. Motions of the two bodies in a direction parallel to the touching surfaces is prevented (or hampered) due to molecular adhesion and/or irregularities on the involved surfaces often too minute to discern with the naked eye.
- **Wet friction** occurs when the surfaces of two solid bodies are not directly in contact but separated by a thin film of lubricants. Again friction will try to hamper the motion but the underlying physics is related to fluid mechanics which is not part of this course.

Friction Laws for Dry Surfaces

Motion or impending motion of two surfaces in contact causes reaction force between them as a frictional force, F. This friction is:

1. Parallel to a flat surface or tangent to a curved surface.
2. Opposite in direction to the motion or impending motion.
3. Dependent of the force pressing the surface together.

4. Independent of the area of surface of contact.
5. Independent of velocity in most cases.
6. Dependent on the nature of the contacting surfaces.

Coefficients of Friction

Consider the case of a block on a horizontal surface being pushed by a force P so that the block has an intending motion to the right (Figure 7.1). Force N, often called the **normal force** (normal = perpendicular), is directed perpendicular to the direction of possible motion (of the block) and can be thought of as the resistance of the surface against penetration by the block. The resistance of course is spread in some fashion over the entire contact area and hence N is a replacement force for this resistance.

The force F is called the **friction force**. It is parallel to the direction of possible motion and acting onto the block in the direction opposing that of possible motion (the motion the block would undertake if friction were absent).

If friction is holding the block in place we know from experience that if the horizontal component of force P is getting too strong the block will ultimately start to slide. The critical case at which sliding is about to set in is referred to as the case of **impending motion** and the value of the friction force in that case is called the **maximum friction force**, $\boldsymbol{F_{max}}$.

Many experiments have been conducted to find out as to what influences the value of F_m. Incidentally nature is kind to us in the case of dry friction. As it turns out F_{max} only depends on the normal force N, the type of materials involved, and the surface roughness. Furthermore, the relationship

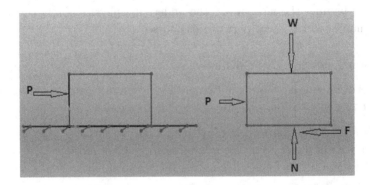

Fig. 7.1 A block on a horizontal surface being pushed and its FBD.

between F_{max} and N is linear for given surfaces:

$$\mu = \frac{F_{\text{max}}}{N}$$

This equation is referred to as the **law of dry friction** and μ_s as the **coefficient of static friction**. Its value is listed below for a variety of combinations of materials for the block and for the supporting surface. Here is a fairly simple experiment to measure the coefficient of static friction.

Experimental Determination of Coefficient of Friction

A block is placed on an inclined plate and the angle of inclination θ is slowly increased from 0 until sliding sets in. The angle at which sliding is impending is referred to as the angle of (static) friction.

The analysis of the FBD of Figure 7.2 (the sum of the force along the incline and the sum of the force perpendicular to that) gives the relationship:

$$\sum F_x = 0 : \quad F_{\text{max}} = W \sin \theta$$
$$\sum F_y = 0 : \quad N = W \cos \theta$$

but

$$\mu = \frac{F_{\text{max}}}{N} = \frac{W \sin \theta}{W \cos \theta} = \tan \theta$$

Therefore, $\mu = \tan \theta$.

Angle of Friction

Friction can be also described in terms of the angle of friction. In this case, the frictional force F_{max} and the normal force N are combined in

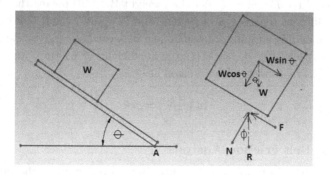

Fig. 7.2 Setup for determining coefficient of friction.

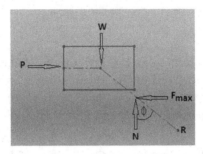

Fig. 7.3 FBD for determining angle of friction.

the resultant R. The frictional angle is the angle between N and R (see Figure 7.3).

$$\tan \phi = \frac{F_{\max}}{N}$$

Belt Friction

Belts are used to turn pulley and drums and have many applications. The two main assumptions relating to belts are:

1. A rope, cable, or flat belt is used (notched belt or a V-belt would require a more involved analysis).
2. Motion is impending, such as the manner of a rope wound around a fixed cylinder so that the rope is starting to slip.

Figure 7.4 shows a simplified case of a rope passed over a fixed cylindrical beam and a force P causing impending motion of the weight upward. Due to friction, force P will be larger than weight W. This means that the rope has two different tensions, a large tension (T_L) and a small tension (T_s).

For impending slipping, the difference between the large tension T_L and the small tension T_s depends on the coefficient of friction, and angle of contact; this relationship is expressed as:

$$\ln \left(\frac{T_L}{T_S} \right) = \mu \theta$$

Rules for Belt Friction Analysis

1. The LARGE tension T_L is generally in the direction of the impending motion.

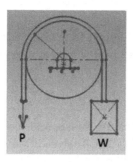

Fig. 7.4 Belt and pulley system.

2. When the pulley is rotating, the SLACK tension T_S is generally in the direction of the rotation.

3. In an inclined plane problem, analyze the forces acting on the body on the inclined plane and accept whichever tension is larger as T_L and whichever is smaller as T_S.

Problem 7.1

A weight, W = 3000 lb is raised by two wedges ($\theta = 7°$) as shown in Figure 7.5. The coefficient of static friction is 0.23 for all surfaces. For an applied force of 1890 lb to the upper wedge, determine the linear displacement of the 3000 lb block upward.

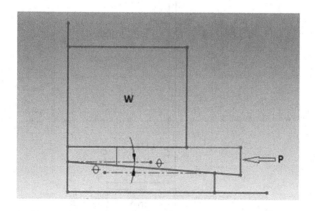

Fig. 7.5 Wedge system.

SolidWorks solution

In this chapter, the steps required to provide SolidWorks solution for problems related to friction are given. The steps are as follows:

1. Create the body.
2. Create the wedge.
3. Create the block.
4. Create the assembly of body, wedge, and block.
5. Create the SolidWork Motion Analysis: Define contacts and apply force to the upper wedge.

Body

The sketch for the body which is extruded 20 inches is shown in Figure 7.6.

Block

The block made from AISI 304 has dimensions: $20 \times 20 \times 25.95$.

Wedge

The wedge (X2) has dimensions: 20×4.33 at the required angle and extruded 25.95.

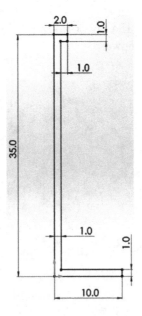

Fig. 7.6 Sketch for the body.

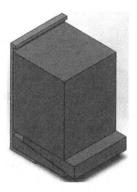

Fig. 7.7 Assembly of body, block, and wedge.

Fig. 7.8 Creating contacts.

Assembly model

The assembly of the body, block, and wedge is shown in Figure 7.7.

Create the SolidWork motion analysis

1. Click the **Contact** tool and under **Selections**, select **Body** and lower **Wedge** (see Figure 7.8).
2. Click the **Contact** tool and under **Selections**, select Lower **Wedge** and upper **Wedge**.
3. Click the **Contact** tool and under **Selections**, select Upper **Wedge** and **Block**.
4. Click the **Force** tool and under **Direction**, select right-face of the upper **Wedge** (see Figure 7.9).
5. Click **Calculate** (see Figure 7.10 for the plot of linear displacement of the block, upward).

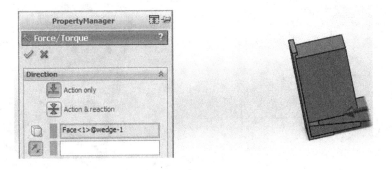

Fig. 7.9 Force PropertyManager.

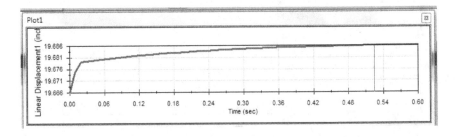

Fig. 7.10 Plot of upward linear displacement of the block.

Summary

This chapter has discussed how SolidWorks is used to handle problems related to friction between contacting bodies. More advanced features of SolidWorks can be used to deal with a rotating body in contact with a stationary body although this is not dealt with here.

Exercises

P1. For an applied force P = 676 N shown in Figure P1, determine the linear displacement of the 100-kg block upward. The coefficient of static friction is 0.15 for all surfaces.

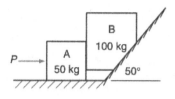

Fig. P1

P2. The minimum force that will pull wedge A to the left is P = 5.2 kN (See Figure P2), determine the linear displacement of the 1500-kg block upward.

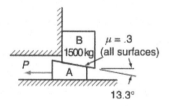

Fig. P2

Reference

Walker, K. M., *Applied Mechanics for Engineering Technology*, Chapter 7, 8th Edition, Prentice Hall, Upper Saddle River, NJ, 2007.

Chapter 8

Centroids and Center of Gravity

Objectives: When you complete this chapter you will have:

- Understanding of how SolidWorks can be applied to determine area moment of inertia of 2D objects encountered in applied mechanics.
- Used SolidWorks to determine area moment of inertia of 2D objects encountered in applied mechanics.

SolidWorks, which is an industry-standard CAD software, seems to be most frequently applied to shape design by many end-users. However, this powerful software has been designed to cater for a number of real-life engineering applications such as:

- Shape design.
- Design of machine elements.
- Applied dynamics/mechanics.
- Engineering (finite element) analysis (FEA).
- Computer-aided manufacturing plug-in (very recent).

We will consider the discussion of SolidWorks applications in applied dynamics, beginning with the determination of centroids, centre of gravity, and area moment of inertia of composite parts. Because SolidWorks is the same software used for shape design by the students, simultaneously using this software for applied dynamics expands its scope of application inside and outside the classroom environment.

Determination of Centroids and Centre of Gravity using SolidWorks

The steps involved in using SolidWorks for determining the centroids and centre of gravity of 2D shapes are very easy:

1. Open a **New SolidWorks Part** Document.

2. Sketch the figure describing the shape.
3. Click the **Section Properties** tool from **SolidWork Command-Manager**.

It is important to follow the specifications given in the orientation of the figure for determining the section properties. For example, if symmetry is required to orient the centroid at the centre of the shape, then it would be useful to include the design intent in the sketching of the shape. If the corner of the shape is to be located at the origin then this should be obeyed. Maintaining the design intents assist greatly in SolidWorks automatically determining the section properties for general shapes. It is that easy and this property of SolidWorks and other CAD software makes computation of section properties very handy for students and designers in industry who may not be interested in using complicated mathematical equations to determine section properties.

Problem 8.1

Determine the centroid of the composite area shown in Figure 8.1.

SolidWorks solution

Open a **New SolidWorks Part** Document.
Sketch Figure 8.1 as given.

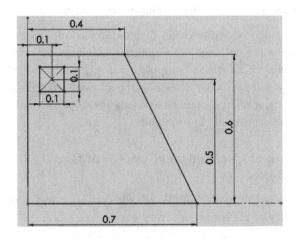

Fig. 8.1 Composite area.

Click the **Section Properties** tool (the section properties are automatically displayed).

Centroid relative to sketch origin (meters): X = 0.287; Y = 0.266 (the solution matches the textbook solution).

Problem 8.2

Determine the moment of inertia about the x-axis for the triangular area shown in Figure 8.2.

SolidWorks solution

Open a **New SolidWorks Part** Document.
Sketch Figure 8.2 as given.
Click the **Section Properties** tool (the section properties are automatically displayed).

Centroid relative to sketch origin (meters): X = 2.08; Y = 5.00.
Area = 18.00 in^2.
Moments of inertia of the area, at the centroid (in^4): Lxx = 81.00; Lyy = 18.45.
$I_x = L_{xx} + Area \times \bar{Y}^2 = 81.00 + 18(5)^2 = 531 \ in^4$ (the solution matches the textbook solution).

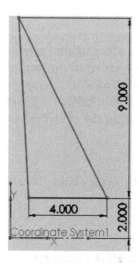

Fig. 8.2 Triangular area.

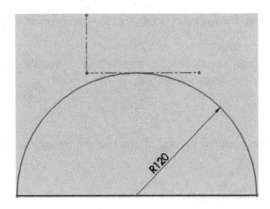

Fig. 8.3 Semi-circular area.

Problem 8.3

Determine the moment of inertia about the x-axis for the semi-circular area shown in Figure 8.3.

SolidWorks solution

Open a **New SolidWorks Part** Document.
Sketch Figure 8.3 as given.
Click the **Section Properties** tool (the section properties are automatically displayed).

Area $= 22619.47 \, \text{mm}^2$.
Centroid relative to sketch origin (millimeters): X $= 49.71$; Y $= -69.07$.
Moments of inertia of the area, at the centroid (millimeters^4): Lxx $= 22759203.36$; Lyy $= 81430081.58$.
$I_x = L_{xx} + Area \times \bar{Y}^2 = 22759203.36 + 22619.47(-69.07)^2 = 130.67 \times 10^6 \, mm^4$ (the solution matches the analytical solution).

Problem 8.4

Determine the moment of inertia about the x-axis for the area shown in Figure 8.4.

SolidWorks solution

Open a **New SolidWorks Part** Document.
Sketch Figure 8.4 as given.

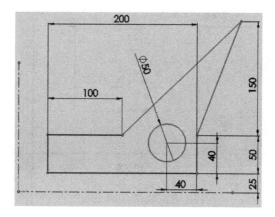

Fig. 8.4 Composite area.

Click the **Section Properties** tool (the section properties are automatically displayed).

Area $= 15536.50 \, \text{mm}^2$.
Centroid relative to sketch origin (meters): X = 173.53; Y = 84.31.
Moments of inertia of the area, at the centroid (millimeters^4):
Lxx = 34608728.41; Lyy = 71703916.75.
$I_x = L_{xx} + Area \times \bar{Y}^2 = 34608728.41 + 15536.50(84.31)^2 = 145 \times 10^6 \, \text{mm}^6$
(the solution matches the analytical solution).

Problem 8.5

Determine the moment of inertia about the horizontal centroidal axis of the composite area shown in Figure 8.5.

SolidWorks solution

Open a **New SolidWorks Part** Document.
Sketch Figure 8.5 as given.
Click the **Section Properties** tool (the section properties are automatically displayed).

Area $= 36.00 \, \text{in}^2$.
Centroid relative to sketch origin (inches): X = 5.00; Y = 2.33.
Moments of inertia of the area, at the centroid (inches^4): Lxx = 108.00; Lyy = 428.00.
$I_x = L_{xx} = 108 \, in^4$ (the solution matches the textbook solution).

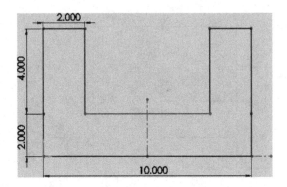

Fig. 8.5 Composite area.

Problem 8.6

For the modified 'C' section shown in Figure 8.6, determine:

1. The coordinates of the centroid (i.e. $\bar{X}$ and $\bar{Y}$). Reference **must** be taken from point 'A'.
2. Moment of inertia about the centroidal x-axis.

SolidWorks solution

Open a **New SolidWorks Part** Document.
Sketch Figure 8.7 in SolidWorks.
Click Section Properties tool (see Figure 8.8) (solutions are automatically displayed).

Problem 8.7

1. Determine the coordinates of the centroid of the area in Figure 8.9.
2. Determine the moment of inertia about the centroidal x-axis.
3. Determine the moment of inertia about the x-axis.

SolidWorks solution

Open a **New SolidWorks Part** Document.
Sketch Figure 8.9 in SolidWorks.
Click Section Properties tool (see Figure 8.10) (solutions are automatically displayed).

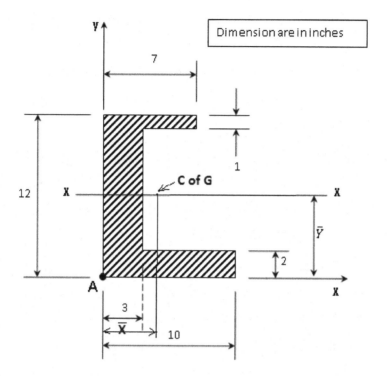

Fig. 8.6 Problem description.

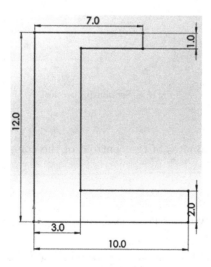

Fig. 8.7 Sketch for problem being solved.

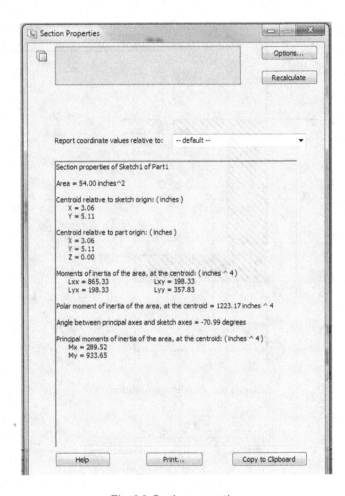

Fig. 8.8 Section properties.

Problem 8.8

Determine the coordinates of the centroid of the composite area shown in Figure 8.11.

SolidWorks solution

Open a **New SolidWorks Part** Document.
Sketch Figure 8.11 in SolidWorks.
Click Section Properties tool (see Figure 8.12) (solutions are automatically displayed).

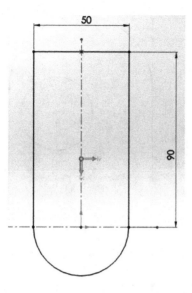

Fig. 8.9 Problem description.

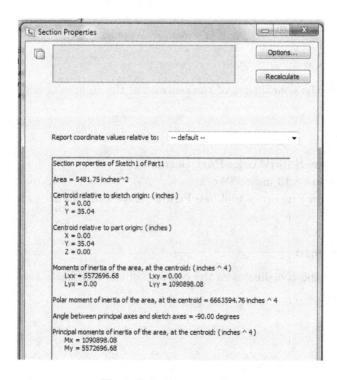

Fig. 8.10 Section properties.

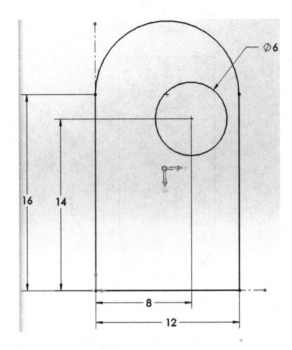

Fig. 8.11 Problem description.

Problem 8.9

Determine the coordinates of the centroid of the composite area shown in Figure 8.13.

SolidWorks solution

Open a **New SolidWorks Part** Document.
Sketch Figure 8.13 in SolidWorks.
Click Section Properties tool (see Figure 8.14) (solutions are automatically displayed).

Problem 8.10

Determine the coordinates of the centroid of the composite area shown in Figure 8.15.

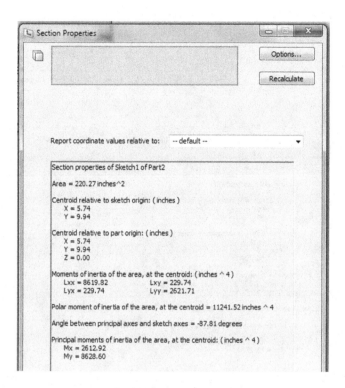

Fig. 8.12 Section properties.

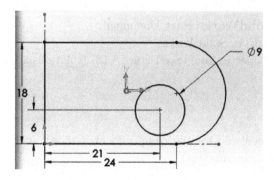

Fig. 8.13 Problem description.

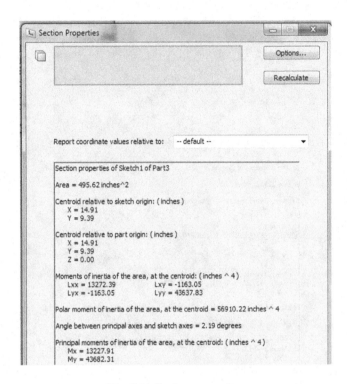

Fig. 8.14 Section properties.

SolidWorks solution

Open a **New SolidWorks Part** Document.
Sketch Figure 8.15 in SolidWorks.
Click Section Properties tool (See Figure 8.16) (solutions are automatically displayed).

Problem 8.11

Determine the centroid for the composite area shown in Figure 8.17 (all dimensions are in millimeters).

SolidWorks solution

Open a **New SolidWorks Part** Document.
Sketch Figure 8.17 in SolidWorks.
Click Section Properties tool (see Figure 8.18) (solutions are automatically displayed).

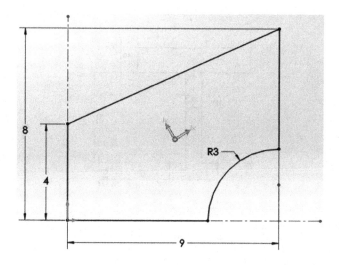

Fig. 8.15 Problem description.

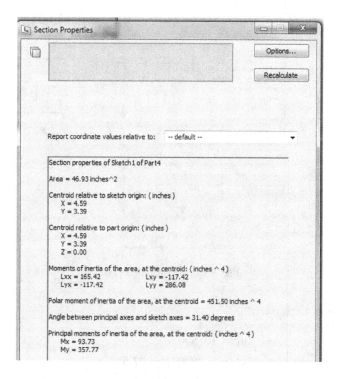

Fig. 8.16 Section properties.

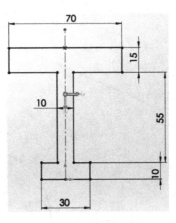

Fig. 8.17 Problem description.

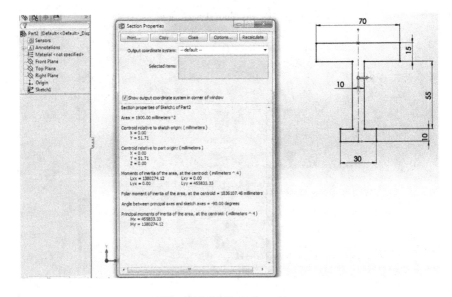

Fig. 8.18 Section properties.

Centroids of each part:

$$y_1 = 10/5 = 5$$
$$y_2 = 55/2 + 10 = 37.5$$
$$y_3 = 15/2 + 55 + 10 = 72.5$$

Part	$A_i\,(\mathrm{mm}^2)$	$y_i\,(\mathrm{mm})$	$A_i\,y_i$
1	$30 \times 10 = 300$	5	1500
2	$55 \times 10 = 550$	37	20,625
3	$15 \times 70 = 1050$	72.5	76,125
	$= 1,900$		98,250

$$\bar{y} = \frac{A_i y_i}{A_T} = \frac{98,250}{1,900} = 51.71\,\mathrm{mm}$$

Problem 8.12

Determine the centroid for the composite area shown in the Figure 8.19 (all dimensions are in millimeters).

Open a **New SolidWorks Part** Document.
Sketch Figure 8.19 in SolidWorks.
Click Section Properties tool (see Figure 8.20) (solutions are automatically displayed).

$$Area = 2902.62\,\mathrm{mm}^2$$

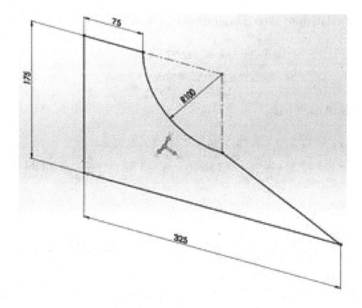

Fig. 8.19 Problem description.

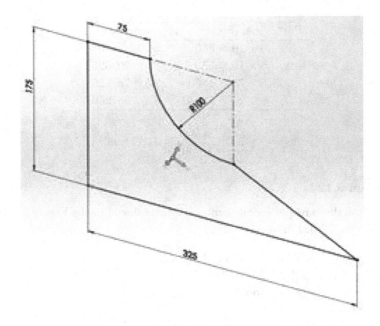

Fig. 8.20 Section properties.

Centroid relative to sketch origin: (millimetres):

$$\bar{x} = 102.27 \, \text{mm}$$
$$\bar{y} = 62.2 \, \text{mm}$$

Centroids of each part:

$$x_1 = 17.5/2 = 8.75; \quad y_1 = 17.5/2 = 8.75$$
$$x_2 = 17.5 - \frac{r}{0.75\pi} = 13.26; \quad y_2 = 17.5 - \frac{r}{0.75\pi} = 13.26$$
$$x_2 = 17.5 - \frac{1}{3}(15) = 22.5; \quad y_2 = 7.5/3 = 2.5$$

Part	A_i (mm^2)	x_i (mm)	$A_i\, x_i$	y_i (mm)	$A_i\, y_i$
1	$(17.5)(17.5) = 306.25$	8.75	2,679.687	8.75	2,679.69
2	$-\pi(10)^2/4 = -78.54$	13.26	$-1,042.766$	13.26	$-1,042.766$
3	$1/2(7.5)(15) = 56.25$	22.5	1,265.625	2.5	140.625
	$= 1,900$		2,902.546		1,777.546

$$\bar{x} = \frac{A_i x_i}{A_T} = \frac{2,902.546}{283.96} = 10.22\,\text{cm} = 102.2\,\text{mm}$$
$$\bar{y} = \frac{A_i y_i}{A_T} = \frac{1,777.546}{283.96} = 6.26\,\text{cm} = 62.2\,\text{mm}$$

Problem 8.13

Determine the moment of inertia about the centroidal x-axis for Figure 8.21 (all dimensions are in millimeters).
Open a **New SolidWorks Part** Document.
Sketch Figure 8.21 in SolidWorks.
Click Section Properties tool (see Figure 8.22) (solutions are automatically displayed).

Problem 8.14

Determine the moment of inertia about the centroidal x-axis for Figure 8.23 (all dimensions are in meters).

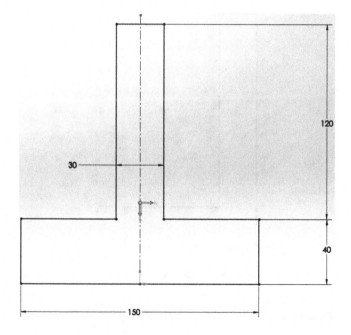

Fig. 8.21 Problem description.

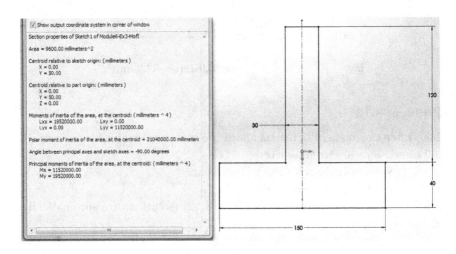

Fig. 8.22 Section properties.

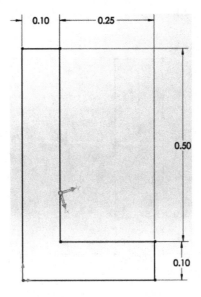

Fig. 8.23 Problem description.

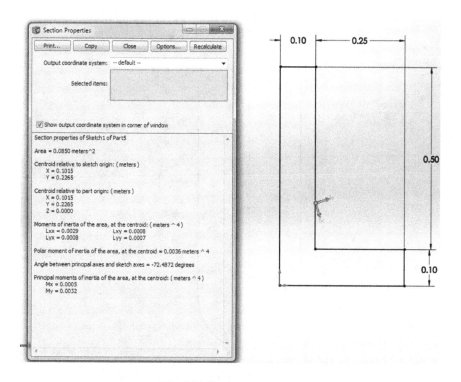

Fig. 8.24 Section properties.

Sketch Figure 8.23 in SolidWorks.
Click Section Properties tool (see Figure 8.24) (solutions are automatically displayed).

Problem 8.15

Determine the moment of inertia about the centroidal x-axis of the area shown in Figure 8.25.
Open a **New SolidWorks Part** Document.
Sketch Figure 8.25 in SolidWorks.
Click Section Properties tool (see Figure 8.26) (solutions are automatically displayed).

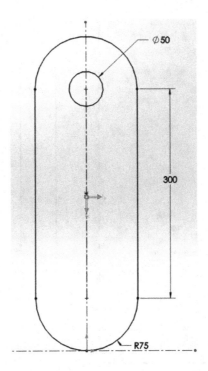

Fig. 8.25 Problem description.

Problem 8.16

A company that produces rackets asks you to determine the moment of inertia of the racket model needed for prototyping about its centroidal x-axis as shown in Figure 8.27. Carry out this task.

Open a **New SolidWorks Part** Document.
Sketch Figure 8.27 in SolidWorks.
Click Section Properties tool (see Figure 8.28) (solutions are automatically displayed).

Problem 8.17

Determine the centroid of the die stamping pattern shown in Figure 8.29.
Open a **New SolidWorks Part** Document.

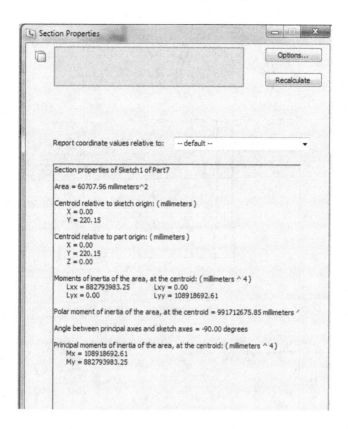

Fig. 8.26 Section properties.

Sketch Figure 8.29 in SolidWorks.

Click Section Properties tool (see Figure 8.30) (solutions are automatically displayed).

Problem 8.18

Determine the centroid of the area shown in Figure 8.31.

Open a **New SolidWorks Part** Document.

Sketch Figure 8.31 in SolidWorks.

Click Section Properties tool (see Figure 8.32) (solutions are automatically displayed).

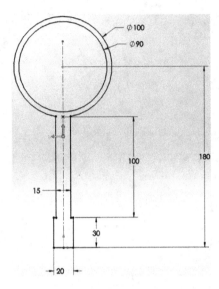

Fig. 8.27 Problem description.

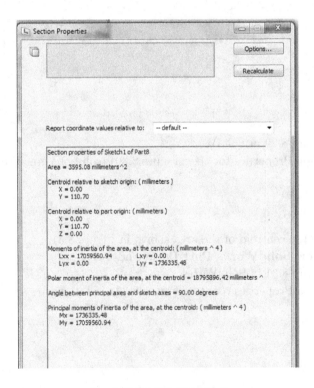

Fig. 8.28 Section properties.

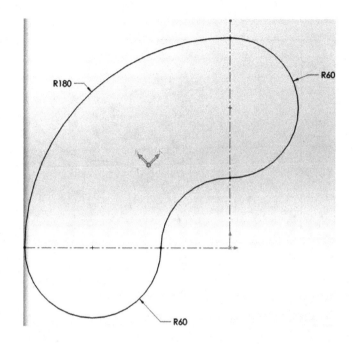

Fig. 8.29 Problem description.

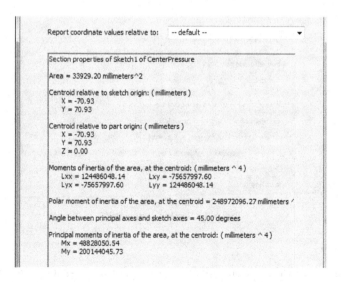

Fig. 8.30 Section properties.

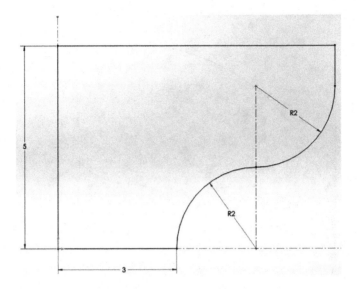

Fig. 8.31 Problem description.

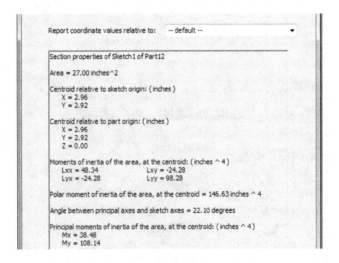

Fig. 8.32 Section properties.

Summary

We have demonstrated that SolidWorks is a Computer-Aided Design Solution Suite that can be applied to a number of mechanical engineering design and analysis applications.

Exercises

P1. Locate the centroid $(\bar{x}, \bar{y})$ of for the area shown in Figure P1.
(Answer: $\bar{x} = 54.40$; $\bar{y} = 11.20$.)

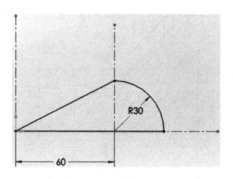

Fig. P1

P2. Locate the centroid $(\bar{x}, \bar{y})$ of for the area shown in Figure P2.
(Answer: $\bar{x} = 48.59$; $\bar{y} = 180.39$.)

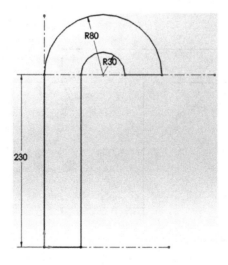

Fig. P2

P3. Locate the centroid $(\bar{x}, \bar{y})$ of for the area shown in Figure P3.
(Answer: $\bar{x} = 3.62$; $\bar{y} = 4.92$.)

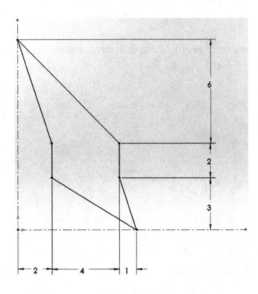

Fig. P3

P4. Locate the centroid $(\bar{x}, \bar{y})$ of for the area shown in Figure P4.
(Answer: $\bar{x} = 5.91$; $\bar{y} = 6.43$.)

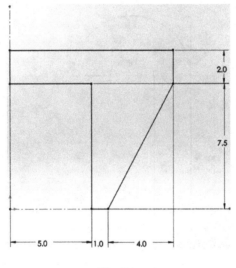

Fig. P4

Reference

Walter K. M., *Applied Mechanics for Engineering Technology*, 6th Edition, Prentice Hall, Pearson Education, Upper Saddle River, NJ, 2000.

Chapter 9

Mass Moment of Inertia

Objectives: When you complete this chapter you will have:

- Understanding of how SolidWorks can be applied to determine moment of inertia of 3D objects of applied mechanics.
- Used SolidWorks to determine moment of inertia of 3D objects of applied mechanics.

SolidWorks seems to be most frequently applied to shape design by many end-users. However, this powerful software has been designed to cater for a number of with real-life engineering applications such as:

- Shape design.
- Design of machine elements.
- Applied dynamics/mechanics.
- Engineering (finite element) analysis (FEA).
- Computer-aided manufacturing plug-in (very recent).

We will consider the discussion of SolidWorks applications in applied dynamics, beginning with the determination of mass moment of inertia of composite parts. Because SolidWorks is the same software used for shape design by the students, simultaneously using this software for applied dynamics expands its scope of application inside and outside the classroom environment.

Determination of Mass Moment of Inertia using SolidWorks

Problem 9.1

In the composite body shown in Figure 9.1, A weighs 25 lb and B weighs 10 lb. Calculate the mass moment of inertia about the z-axis.

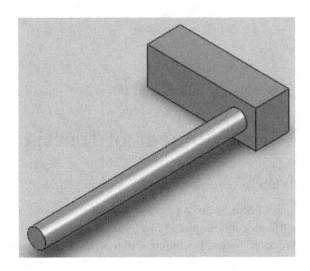

Fig. 9.1 Composite body of A and B.

SolidWorks solution

Density $= 247.21\,\text{lb/ft}^3$.
Mass $= 35.03\,\text{lb}$.
Volume $= 0.14\,\text{ft}^3$.
Surface area $= 2.45\,\text{ft}^2$.
Center of mass (feet): X $= 0.60$; Y $= 0.17$; Z $= 0.31$.

Moments of inertia (pounds per square feet), taken at the center of mass and aligned with the output coordinate system.

Lxx $= 11.68$;   Lyy $= 14.33$;   Lzz $= 3.14$.

$$I_z = L_{zz} + \text{Mass} \times \bar{X}^2 = \frac{1}{32}[3.14 + 35.03 \times (0.6)^2] = 0.49\,\text{ft} - \text{lb} - \text{s}^2.$$

(The solution matches the textbook solution with slight variation: 0.49 and 0.482, respectively. See explanation below.)

Comments: This example is interesting because the reader should understand that different materials must be tested to ensure that we get the right weights indicated in the problem description since SolidWorks uses actual materials. Also, it should be noted that the unit of moments of inertia in SolidWorks is (pounds per square feet), whereas the textbook uses a different unit: ft-lb-s^2 (dividing by g which is gravity acceleration of value $= 32$).

Hence, we divided our solution by 32 to covert our solution to the same unit as that for the textbook.

Problem 9.2

Calculate the mass moment of inertia about the x-axis for the thin rectangular plate shown in Figure 9.2. The plate had a mass of 5 kg before the hole material (1.07 kg) was cut out.

SolidWorks solution

In our SolidWorks model, the z-axis is normal to the plate, so we pay attention to this direction in our solution.

$$I = L_{zz} = 2.0229 \, \text{kg/m}^2.$$

(The solution matches the textbook solution with a slight variation: 2.0229 and 2.1, respectively. See explanation below.)

Comments: This example is again interesting, similar to the preceding one. The problem description specifies 5 kg mass for the plate before the hole material (1.07 kg) was cut out. In SolidWorks approach, we must first create a 5 kg mass based on the given dimension and experimentation with varying thickness and materials until we create the given mass. When this

Fig. 9.2 Thin rectangular plate.

is achieved, we do not need to bother about the hole material (1.07 kg) cut out because this is automatically taken care of once we include the cut hole as shown in Figure 8.21.

Problem 9.3

A steel forging stocked and utilized in a production department of a company consists of a $9 \times 3 \times 3$ in rectangular prism and two cylinders of radius 1.5 in and length 6 in as shown in Figure 9.3. Each cylinder is offset 3 in from the x-axis. Determine the moment of inertia of the forging with respect to the x-coordinate axis given that the specific weight of steel is $490\,\text{lb/ft}^3$.

SolidWorks solution

Open New SolidWorks Part Document.
Sketch a 3×3 in rectangle.
Extrude through a distance of 9 (ensure symmetry).
Sketch a 1.5 diameter circle.
Extrude through a distance of 6.

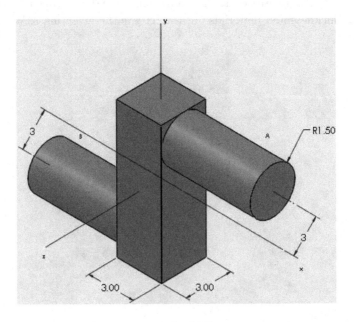

Fig. 9.3 Steel billet.

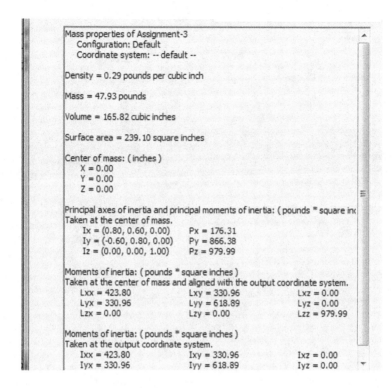

Fig. 9.4 Mass properties for steel billet.

Open New SolidWorks Assembly Document.

Insert the prism.

Insert the cylinder and apply Mating conditions (lower cylinder).

Insert the cylinder again and apply Mating conditions (upper cylinder).

Click Mass Properties tool (see Figure 9.4).

Summary

We have demonstrated that SolidWorks is a Computer-Aided Design Solution Suite that can be applied to a number of mechanical engineering design and analysis applications.

Reference

Walter, K. M., *Applied Mechanics for Engineering Technology*, 6th Edition, Prentice Hall, Pearson Education, Upper Saddle River, NJ, 2000.

PART 2
Dynamics

Chapter 10

Kinematics: Rectilinear Motion Using SolidWorks

Objectives: When you complete this chapter you will:

- Have understood the difference between distance and displacement, speed, and velocity.
- Solve for displacement, velocity, and acceleration using the three equations of constant acceleration linear kinematics.
- Solve problems relating to projectile motion.
- Have understood the concept of using SolidWorks to solve kinematics–rectilinear motion problems.

Introduction

Kinematics is the study of the geometry of motion without considering the forces that are causing the motion. It is concerned with quantities such as displacement, distance, velocity, and acceleration. There are several types of planar motions:

1. *Rectilinear or translational motion.* A particle moves in a straight line, without rotation.
2. *Circular motion.* A particle moves in a circular motion due to rotation.
3. *General plane motion.* A particle that may have both rectilinear and rotational motion simultaneously, as in the case of the link of slider crank mechanism.

What is rectilinear or translational kinematics?

Rectilinear (or translational) kinematics is the linear movement of a body without considering the forces that is causing the motion. Linear motion of a body is also called the rectilinear motion and is analyzed in terms of displacement, distance, velocity, and acceleration.

Linear Displacement

Displacement and distance are two different quantities. While distance is a scalar quantity, displacement is a vector quantity which has both magnitude and direction. Linear displacement is a vector quantity while linear distance is scalar quantity. The symbol used for linear displacement is s. s is measured either in meters, feet, or kilometers.

Linear Velocity

Linear velocity is the rate of change of linear displacement. It is represented by v. It is defined as $v = \frac{s}{t} = \frac{\Delta s}{\Delta t}$, where v is the average angular velocity, s is the linear displacement, and t is the time. The unit of v is m/s, km/h, ft/s, or ft/min and the direction must be given to complete the definition of velocity.

Linear Acceleration

Linear acceleration is the rate of change of linear velocity. It is represented by $\mathbf{a}$ and is defined as $\mathbf{a} = \frac{v}{t} = \frac{\Delta v}{\Delta t}$, where $\mathbf{a}$ is the average linear acceleration, v is the linear velocity, and t is the time. The unit of is $\mathbf{a}$ is m/s^2, mm/s^2, in/s^2, or ft/s^2.

Problem 10.1

An object is released from a 2000 m elevation, during which it falls freely for 8 seconds. The object then decelerates for 5 seconds to a velocity of 7 m/s, which it maintains until it lands. Complete the following:

1. Show a clear diagram of the object's motion.
2. Calculate the maximum velocity of the object.
3. Calculate the total time elapsed during the travel (i.e. from release until the object hits the ground).

SolidWorks Solution

1. The motion diagram is shown in Figure 10.1.
2. The velocity at point B is first obtained as:

$$v_B = v_{A(=0)} + 9.81(8) = 78.48 \text{ m/s}$$

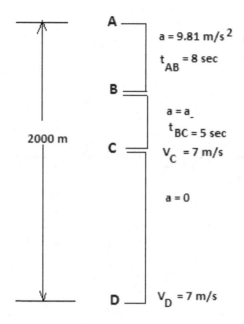

Fig. 10.1 Motion diagram for Problem 10.1.

3. Velocity diagram using SolidWorks:

The velocity at point A is 0, so AB is sketched to have a 'rise' of 78.48 while BC is sketched to stop at a velocity of 7 as shown in Figure 10.2. CD is drawn as a horizontal line but we do not know where point D stops. Its position is adjusted while we check for the section properties until the area under the curve is 2000, which is the given displacement. At this juncture we have to find the length of CD which gives the time required for the third segment of the motion.

Click **Evaluate > Section Properties**

The Section Properties manager pops up as shown in Figure 10.3.

By adjusting the right vertical edge of the shape (see Figure 10.4), such that the area under the curve is 2000 (as given in the problem description), the value of the time for the third segment is obtained as 210.357 seconds. This value closely agrees with the solution of 210.34 seconds.

Verification of solution

$$A_1 = \frac{1}{2}(98)(78.48) = 313.92 \text{ m}^2$$

$$A_2 = \frac{1}{2}(78.48 + 7)(5) = 213.7 \text{ m}^2$$

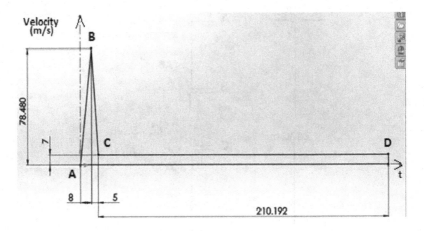

Fig. 10.2 Velocity diagram using SolidWorks.

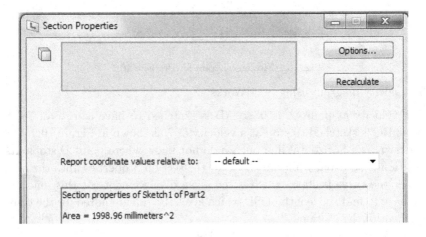

Fig. 10.3 Section Properties manager.

$$A_3 = 7t_3 \, \text{m}^2$$
$$A_1 + A_2 + A_3 = 2000$$
$$313.92 + 213.7 + 7t_3 = 2000$$
Solving, leads to: $t_3 = 210.34 \, \text{s}$
$$t_1 + t_2 + t_3 = 8 + 5 + 210.34 = 223.34 \, \text{s}$$

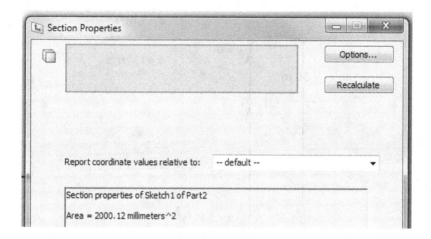

Fig. 10.4 Final area obtained by adjusting shape.

Problem 10.2

A helicopter accelerates uniformly upward at $1\,\text{m/s}^2$ to a height of $300\,\text{m}$. By the time it reaches $350\,\text{m}$, it has decelerated to 0 vertical velocity. It then accelerates horizontally at $4\,\text{m/s}^2$ to a velocity of $15\,\text{m/s}$. Determine the total time required for this sequence.

SolidWorks solution

The motion diagram is shown in Figure 10.5.

Velocity diagram using SolidWorks

There are three sections of motion for this problem. For the first section, the distance covered is $300\,\text{m}$ (given).

$$300 = v_{o\,(=0)} + \frac{1}{2}(1)t^2; \quad => t = \sqrt{2(300)} = 24.5\,\text{s}$$

The velocity at point A is 0 and the time taken for the helicopter to accelerate to the height of $300\,\text{m}$ is $24.5\,\text{s}$.

1. **Sketch** a triangle so that the time, **AC = 24.5** (see Figure 10.6).
2. **Stretch** line AB until the area of triangle ABC = 300.
3. Each time AB is stretched click **Evaluate > Section Properties.** (Continue until the area of ABC = 300.)

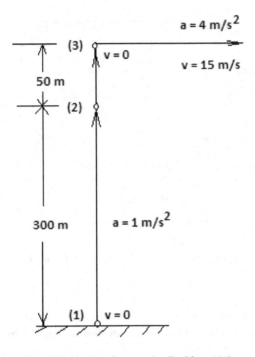

Fig. 10.5 Motion diagram for Problem 10.2.

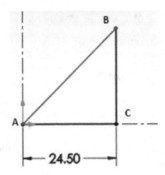

Fig. 10.6 Partial solution to the first section of the motion.

For the second section, the helicopter has reached 350 m, so that the distance covered is 50 m (350–300); it also decelerated to 0.

4. **Sketch** a triangle BCD, then delete line **BC** (see Figure 10.7). (Solid-Works will not handle multiple areas.)

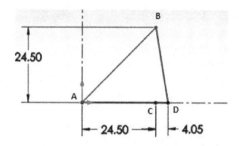

Fig. 10.7 Partial solution to the second section of the motion.

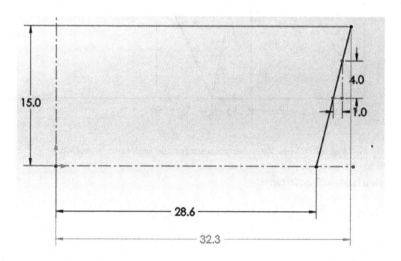

Fig. 10.8 Partial solution to the third section of the motion.

5. Adjust the length of CD until the total area of ABD = **350** (see Figure 10.7).
6. Measure the distance of CD (= 4.05).

   In this problem, we cannot include the third section of the travel on the same figure because SolidWorks will not handle multiple areas so this is done separately (see Figure 10.8).

   The total time required for this sequence of motion is 32.3 s as shown in Figure 10.8.

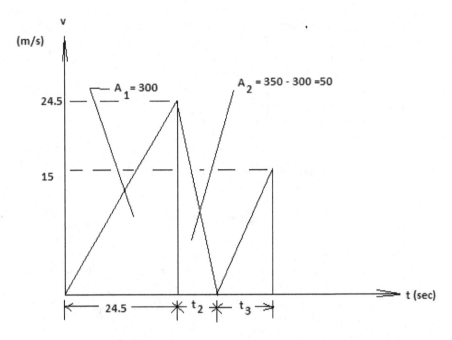

Fig. 10.9 Velocity–time diagram for the helicopter's motion.

Verification of solution

For the first section, the distance covered is 300 m (given).

$$300 = v_{o\,(=0)} + \frac{1}{2}(1)t^2; \quad => t = \sqrt{2(300)} = 24.5\,\text{s}$$

The velocity–time diagram for the helicopter's motion is shown in Figure 10.9.

$$A_1 = \frac{1}{2}(v)(24.5) = 300\,m^2; \quad v = 2(300)/24.5 = 24.5\,\text{m/s}$$

$$A_2 = \frac{1}{2}(24.5)(t_2) = 50\,\text{m}^2; \quad t_2 = 2(50)/24.5 = 4.08$$

$$15 = v_{o\,(=0)} + 4t_3; \quad => t_3 = 15/4 = 3.75\,\text{s}$$

Solving leads to: $t_1 + t_2 + t_3 = 24.5 + 4.08 + 3.75 = 32.33\,\text{s}$.

Problem 10.3

A car decelerates uniformly from an initial velocity of 8 m/s (point A) to a final velocity of 2 m/s (point B). This velocity is maintained for 20 s to reach point C. It then brakes uniformly at $0.8\,\mathrm{m/s^2}$ to come to rest at point D. If the total displacement from A to D is 100 m, determine:

1. The total time from A to D.
2. The acceleration rate from A to B.

SolidWorks solution

1. Sketch line AB such that OA = 8 and the vertical height of B from OD is 2.
2. Sketch BC = 20 in length (time taken; see Figure 10.10).
3. Sketch CD with rise = 0.8 and fall = 1.
4. Sketch OD to close the area.
5. Move the point B horizontally, each time click **Evaluate > Section Properties**, until the area under the curve = 100.
6. Dimension the times take from A to B (11.55 or 11.6) and C to D (2.5).

The total time $= 11.55 + 20 + 2.5 = 34.05$ s.

Problem 10.4

An object travels from A to B on the portable conveyor (length, $L = 6\,m$; rise $dy = 4$ and fall $dx = 5$), shown in Figure 10.11. The conveyor belt speed is 1.5 m/s. The complete portable conveyor is traveling at $7.2\,\mathrm{km/h}$ to the left. Determine the total displacement of the object as it travels from A to B.

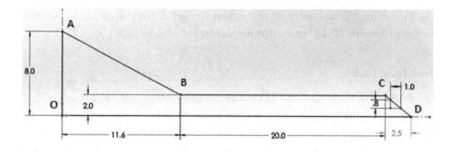

Fig. 10.10 Velocity diagram using SolidWorks.

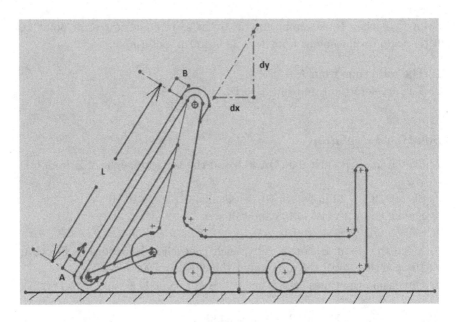

Fig. 10.11 Problem geometric definition.

SolidWorks solution

Displacement from A to B on conveyor = 6 m.

　　Conveyor belt speed = 1.5 m/s.

$$t = \frac{\text{displacement}}{\text{speed}} = \frac{6}{1.5} = 4 \, \text{s}.$$

Complete portable conveyor is traveling at 7.2 km/h $\left(\frac{7200 \, \text{m}}{60 \times 60 \, \text{s}} = 2 \, \text{s}\right)$.

　　For the time of 4 seconds, the displacement is $\Delta s = \Delta v \times \Delta t = 2 \times 4 = 8$ m; using the values of displacement to sketch the displacement vectors shown in Figure 10.12, the total displacement = 5 m at 48.5°.

Problem 10.5

An object travels from A to B on the portable conveyor (length, $L = 8$ m; rise $dy = 1$ and fall $dx = 1$) shown in Figure 10.11. The conveyor belt speed is 2.0 m/s. The complete portable conveyor is traveling at 8.0 km/h to the left. Determine the total displacement of the object as it travels from A to B.

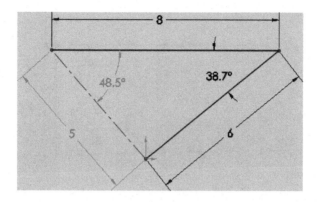

Fig. 10.12 Displacement vectors.

SolidWorks solution

Displacement from A to B on conveyor = 8 m.
Conveyor belt speed = 2 m/s.

$$t = \frac{\text{displacement}}{\text{speed}} = \frac{8}{2} = 4\,\text{s}.$$

Complete portable conveyor is traveling at 8 km/h $\left(\frac{8000\,\text{m}}{60\times 60\,\text{s}} = 2.22\,\text{s}\right)$.

For the time of 4 seconds, the displacement is $\Delta s = \Delta v \times \Delta t = 2.22 \times 4 = 8.88\,m$; using the values of displacement to sketch the displacement vectors shown in Figure 10.13, the total displacement = 6.51 m at 60° to the horizontal.

Results verification

Problem 10.4: Using cosine rule for Figure 10.12:
$s^2 = 8^2 + 6^2 - 2(8)(6)(\cos 38.7°) = 25.08$; $s = 5\,\text{m}$

Using sine rule for Figure 10.12 : $\frac{5}{\sin(38.7)} = \frac{6}{\sin\alpha}$; $\quad \alpha = 48.6°$

Problem 10.5: Use cosine rule for Figure 10.12:
$s^2 = 8^2 + 8.88^2 - 2(8)(8.88)(\cos 45°) = 42.38$; $s = 6.51\,\text{m}$

Using sine rule for Figure 10.12 : $\frac{6.51}{\sin(45)} = \frac{8}{\sin\alpha}$; $\quad \alpha = 60.3°$

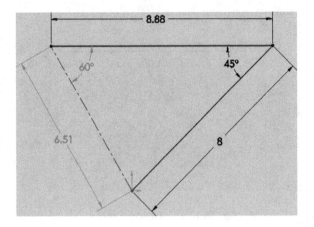

Fig. 10.13 Displacement vectors.

Projectiles

A projectile traces a path (trajectory) during its travel. Projectile motion consists of two rectilinear motions occurring simultaneously (i.e. vertical and horizontal). Each motion can be treated separately or added together as vectors. Both vertical and horizontal motions can be represented by:

- Displacement vectors.
- Velocity vectors.
- Acceleration vectors.

Assume 0 air resistance.
Therefore, the only factors affecting projectile motion:

- Initial velocity.
- Projectile's direction.
- Gravity (g).

Acceleration is constant in both directions:

- horizontally (i.e. $v_i = v_f$).
- $9.81 \, \text{m/s}^2$ or $32.2 \, \text{ft/s}^2$ vertically.

(Note: Generally, gravity is negative for projectiles moving upwards and gravity is positive for projectiles moving downwards.)

Problem 10.6

A 1.67-ft diameter (spherical) ball is dropped from the top of a house which is 8.33 ft high onto a pavement. The initial velocity of the ball is $V_{ox} = 12.5$ ft/s. Determine the time the ball hits the ground the first time.

SolidWorks solution

1. Model the ball, which is a sphere of 20 in diameter (change units to inches) (Figure 10.14).
2. Model pavement (ground) 30×500 (units in inches) (Figure 10.15).
3. Assemble pavement (ground) and ball (sphere) (Figure 10.16).

Three Mates are defined in the assembly:

• **Top Planes** of ball and ground are **100 in** apart (Figure 10.17).

Fig. 10.14 Ball modeled as a sphere.

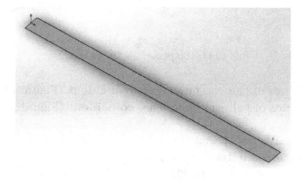

Fig. 10.15 Pavement modeled as the ground.

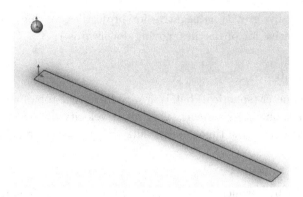

Fig. 10.16 Assembly of pavement (ground) and ball (sphere).

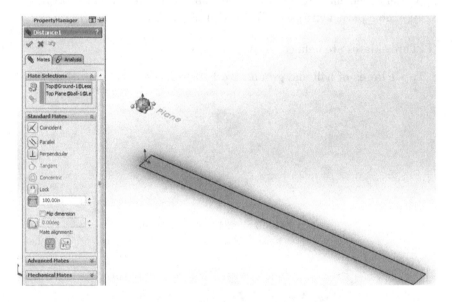

Fig. 10.17 Top planes are 100 in apart.

- **Right Planes** of ball and ground are **coincident** (Figure 10.18).
- **Front Planes** of ball and ground are **coincident** (Figure 10.19).

4. Add-In SolidWorks Motion (Figure 10.20).
5. Select **Motion Analysis** (Figure 10.21)
6. Define **Contact** (Figure 10.22).

    Both the ball and ground are selected when the **Contact** tool is clicked.

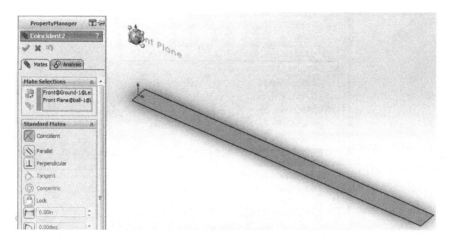

Fig. 10.18 Front planes are coincident.

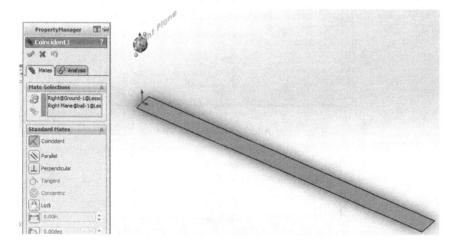

Fig. 10.19 Right planes are coincident.

7. Define **Gravity** (Figure 10.22).
   Click **Gravity** tool and choose *downward* direction.
8. Select **Velocity** (Figure 10.23).
   Right-click the ball in the Motion Analysis Manager and click the Initial Velocity option to define the velocity, $V_{ox} = 150\,\text{in/s}$. Choose the edge of the ground that is longest as the direction.

Fig. 10.20 Add-In SolidWorks Motion.

Fig. 10.21 Select Motion Analysis option.

Fig. 10.22 Gravity and contact definitions.

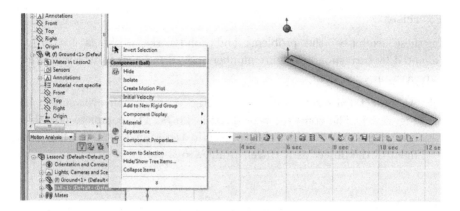

Fig. 10.23 Select Velocity.

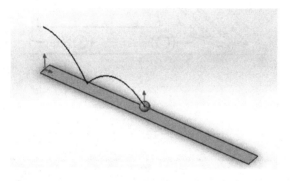

Fig. 10.24 Simulation of Velocity.

9. Click **Calculate** to **Simulate** the motion with time of 1.8 seconds (Figure 10.24).

Summary

In this chapter, rectilinear motion of machine parts have been presented in such a way that students can appreciate the application of SolidWork to solve this class of kinematic problems (including projectile motion). This chapter contains some tutorials at the end, with the solutions given. Students should go through the tutorial solution to see how the principles presented in this chapter are applied. SolidWorks solutions have been verified to agree with solutions of theoretical methods.

Exercises

In these exercises, the problems and SolidWorks workings are given (denoted by corresponding figure number and '-S') and you are to fill in the gaps as your exercises.

P1. An object travels from A to B on the portable conveyor shown in Figure P1. The conveyor belt speed is 1 m/s. The complete portable conveyor is traveling at 9 m/s to the left. Determine the total displacement of the object as it travels from A to B.

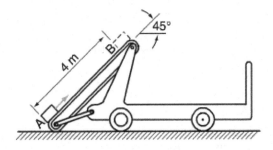

Fig. P1

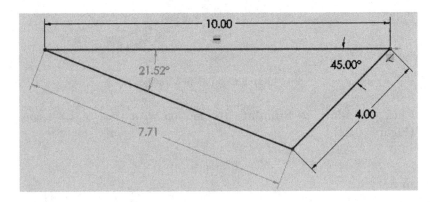

Fig. P1-S

P2. A forklift is initially at the position shown in Figure P2. Determine the displacement of the forks when the boom is extended 16 ft and lifted from 12° to 35°.

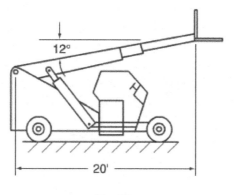

Fig. P2

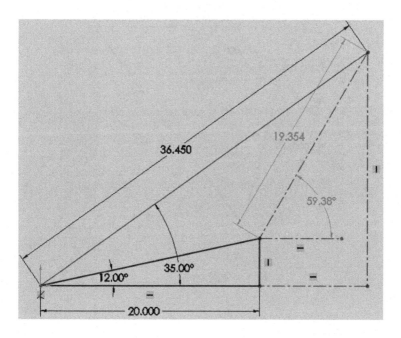

Fig. P2-S

P3. A car is traveling at a constant speed of 90 km/h takes 8 seconds to travel through a constant radius curve shown in Figure P3. Determine the average acceleration of the car.

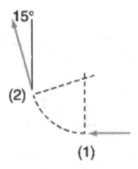

Fig. P3

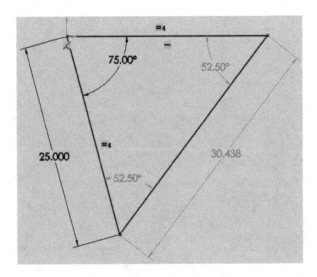

Fig. P3-S

P4. The object shown in Figure P4 moves from point (1) to point (2) in a circular path at a constant speed of 30 m/s in a time of 4 seconds. Determine the average acceleration of the car.

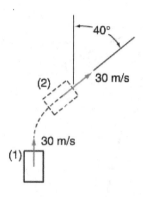

Fig. P4

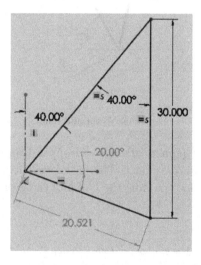

Fig. P4-S

P5. Study how vectors are added as shown in Figure P5.

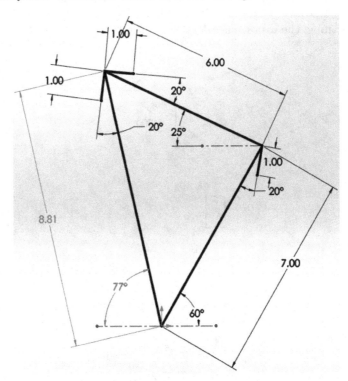

Fig. P5-S Vector addition.

P6. A 12 in diameter (spherical) ball is dropped from the top of a house, 120 in high onto a pavement. The initial velocity of the ball is $V_{ox} = 120$ in/s. Determine the time the ball hits the ground the first time.

References

Walter, K. M., *Applied Mechanics for Engineering Technology*, 6th Edition, Prentice Hall, Pearson Education, Upper Saddle River, NJ, 2000.

Chang, K.-H., *Motion Simulation and Mechanism Design with SolidWorks Motion*, SDC, Mission, KS, 2009.

Chapter 11

Kinematics: Angular Motion

Objectives: When you complete this chapter you will have:

- Understanding of how SolidWorks can be applied to determine angular displacement, velocity, and acceleration.
- Distinguished between radial acceleration, tangential acceleration, and total acceleration.

Introduction

The rotor of an electric motor, the rotor carrying turbine blades, and the rotor carrying the blades of a ceiling fan are examples of elements that have rotational or angular motion. There are similarities between linear and angular motion in terms of the displacement, velocity, and acceleration. The only difference is that in one case these parameters are linear, while in the other case they are angular.

Angular Displacement

When the lever AB rotates (Figure 11.1), the tip B travels through $2\pi r$ in one complete revolution. When AB turns through a distance equal to the AB on the circumference then the angle turned is 1 radian. This means that there are 2π radians in one revolution. The radian is a dimensionless unit. The relationships used in angular displacement θ are:

$$\text{One revolution} = 360° = 2\pi \text{ radian.}$$

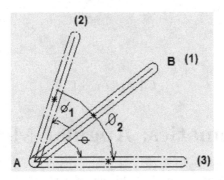

Fig. 11.1 Angular displacement of AB.

Angular Velocity

Angular velocity is the rate of change of angular displacement with respect to time. Mathematically, angular velocity is defined as:

$$\omega = \frac{\Delta\theta}{\Delta t}$$

The unit for angular velocity is rad/s. In some cases, a non-standard unit used is the revolution per minute (rpm). If angular velocity is given in N-rpm, then the conversion to radians is given by:

$$\omega = \frac{2\pi N}{60}$$

Angular Acceleration

Angular acceleration is the rate of change of angular velocity with respect to time. Mathematically, angular acceleration is defined as:

$$\alpha = \frac{\Delta\omega}{\Delta t}$$

The unit for angular velocity is rad/s^2.

Relationship Between Linear and Angular Motion

The relationship between linear and angular motion is easily understood by considering a rope wound around a pulley with one end dangling with a weight as shown in Figure 11.2.

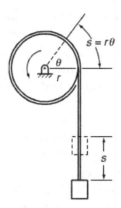

Fig. 11.2 Relating linear displacement to angular displacement.

When the weight on the dangling side move a distance s, then the arc length which the rope winds on the pulley is given as:

$$s = r\theta$$

From the previous discussion the velocity is obtained by differentiating the displacement with respect to time:

$$v = \frac{ds}{dt} = r\frac{d\theta}{dt} = r\omega$$

and the acceleration is obtained by differentiating the velocity with respect to time:

$$a = \frac{dv}{dt} = r\frac{d\omega}{dt} = r\alpha$$

Therefore, we can summarize the relationships between linear motion and angular motion as follows:

$$s \to \theta : s = r\theta$$
$$v \to \omega : v = r\omega$$
$$a \to \alpha : a = r\alpha$$

The connection between linear motion and angular motion is the radius of rotation, r of a body.

Radial and Tangential Acceleration

When a string is wound around our finger at one end and the other end is attached a weight, if the string is rotated, we will observed some tension in our finger. This is because the weight produces a centripetal acceleration toward our finger and since the weight has a mass there is an accompanying centripetal force toward our finger. In short, we can say that there are two types of accelerations in general: Tangential and radial. Tangential acceleration is the one that we have already discussed. Radial acceleration is normal to tangential acceleration, acting toward the centre of rotation. Applying vector mathematics, both accelerations can be summed to produce the total acceleration.

There are other accelerations which are not covered here: Sliding and Coriolis accelerations. When a rod is rotating about one of the ends, and it carries a sleeve that can slide along the rod, this leads to radial acceleration on the sleeve. There is the Coriolis acceleration which is normal the sliding acceleration. In general therefore, we can have as much as four components of acceleration: Radial and sling along the radius of the radius, tangential and Coriolis perpendicular to the circular path at any instant.

Angular Motion with Uniform Acceleration

The three basic equations for linear motion with uniform acceleration are similar to the angular motion with uniform acceleration, with the appropriate transformations as follows:

$$s = v_o t + \frac{1}{2}at^2 \qquad \theta = \omega_o t + \frac{1}{2}\alpha t^2$$

$$v = v_o + at \qquad \omega = \omega_o + \alpha t$$

$$v^2 = v_o^2 + 2as \qquad \omega^2 = \omega_o^2 + 2\alpha\theta$$

Problem 11.1

A flywheel with a 32 in diameter, weighing 500 lb, and initially at rest accelerates to 150 rpm in 3 minutes. The flywheel maintains this rpm for 10 minutes. It then decelerates to a complete stop in 7 minutes. Complete the following:

1. Sketch the complete cycle clearly showing critical points.
2. Calculate the angular acceleration and deceleration.
3. Find the total number of revolutions completed during the cycle.
4. Find the total acceleration at the outside diameter of the flywheel 15 minutes into the cycle.
5. Find the kinetic energy at 15 minutes.

SolidWorks solution

$$\omega_o = \frac{2\pi(0)}{60} = 0 \qquad \omega = \frac{2\pi(150)}{60} = 15.7$$

$$t_1 = 180\,\text{s}\,(3\text{ min}); \quad t_2 = 600\,\text{s}\,(10\text{ min})\,t_3 = 420\,\text{s}\,(7\text{ min})$$

1. Sketch the velocity–time diagram using SolidWorks (see Figure 11.3); this involves the following steps:
   Sketch AB = 180 (t_1).
   Sketch line BC = 600 ($t_2 = 2$ minutes).
   Sketch CD = 420 (t_3).
   Sketch line FA; height = 15.7 (final angular velocity).
   Sketch line FG; constant at height = 15.7 (final angular velocity).
   Sketch line GD; height = 15.7 (final angular velocity).
   Sketch AE = 900 (15 minutes).
2. Angular acceleration = FB/AB = 15.7/180 = 0.087 rad/s^2.
   Acceleration deceleration = GC/CD = 15.7/420 = 0.037 rad/s^2.
3. Click the Mass Properties tool for SolidWorks to calculate the area under the curve = 14,130 rad.
   Convert this to revolution = $14,130/(2\pi) = 2248.86$ rev.
4. Total acceleration at the outside diameter of the flywheel 15 minutes (900 s) into the cycle:
   From E sketch the line HE; measure it. HE = 11.214 rad/s.
   Use the following equations to determine the total acceleration at the outer radius = 16 in = 16/12 = 1.333 ft:

$$a_r = 1.333(11.214)^2 = 167.63\,\text{ft/s}^2$$

$$a_t = 1.333(0.037) = 0.049\,\text{ft/s}^2$$

Total acceleration = $\sqrt{a_r^2 + a_t^2} = \sqrt{167.63^2 + 0.049^2} = 167.63$ ft/s^2.

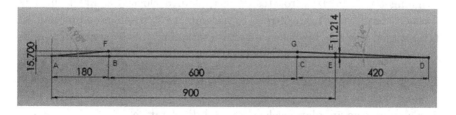

Fig. 11.3 Velocity–time diagram (complete cycle).

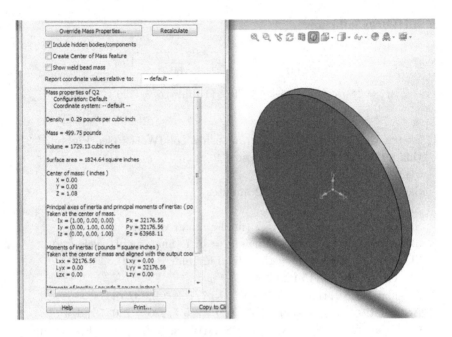

Fig. 11.4 Mass properties.

5. SolidWorks is used to determine the mass moment of inertia of the flywheel (Figure 11.4):

$$I_{zz} = 63968.11\,\text{lb} - \text{in}^2 = \frac{63968.11}{32.2(12^2)} = 13.79\,\text{lb} - \text{ft} - /\text{s}^2.$$

$$KE = \frac{1}{2}I_{zz}\omega^2 = \frac{1}{2}(13.79)(11.21)^2 = 872.67\,\text{ft} - \text{lb}.$$

Problem 11.2

A flywheel with a 64 in outside diameter, an inside diameter of 56 in, and weighing 500 lb initially rotates at 60 rpm in a clockwise direction. The flywheel decelerates from 60 rpm to a complete stop while completing 40 revolutions. The flywheel then remains stationary (i.e. '0' rpm) for 2 minutes. It then accelerates to 95 rpm in 50 revolutions. Complete the following:

1. Sketch the complete cycle clearly showing critical points (clockwise rotation must be taken as positive 'y').
2. Calculate the angular acceleration and deceleration.
3. Find the total time to complete the cycle.

4. Find the total acceleration at the outside diameter of the flywheel 4 minutes into the cycle.
5. Find the kinetic energy 4 minutes into the cycle.

SolidWorks solution

$$\omega_o = \frac{2\pi(60)}{60} = 6.28. \qquad \omega = \frac{2\pi(95)}{60} = 9.95.$$

$$\frac{1}{2}(t_1)(6.28) = 40\,\text{rev} = 40(2\pi) = 251.32\,\text{rad}$$

$$\therefore\ t_1 = 2(251.32)/6.28 = 80\,\text{s}$$

$$\frac{1}{2}(t_3)(9.95) = 50\,\text{rev} = 50(2\pi) = 314.16\,\text{rad}$$

$$\therefore\ t_3 = 2(314.16)/9.95 = 63.16\,\text{s}$$

1. Sketch the velocity–time diagram using SolidWorks (see Figure 11.5); this involves the following steps:
   Sketch line FA = 6.28 (initial angular velocity).
   Sketch AB = 80 (t_1).
   Sketch line BC = 120 (t_2 = 2 minutes).
   Sketch CD = 63.16 (t_3).
   Sketch line GD = 9.95 (final angular velocity).
   Sketch AE = 240 (4 minutes).
2. Angular acceleration = $-AF/AB = -6.28/80 = -0.079$ rad/s^2.
   Acceleration deceleration = DG/CD = 9.95/63.16 = 0.157 rad/s^2.
3. Total time = $t_1 + t_2 + t_3 = 80 + 120 + 63.16 = 263.16$ s.
4. Total acceleration at the outside diameter of the flywheel 4 minutes (240 s) into the cycle: From E sketch the line HE; measure it. HE =

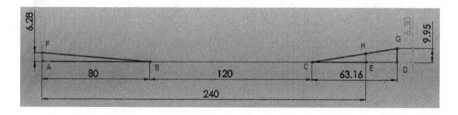

Fig. 11.5 Velocity–time diagram (complete cycle).

6.3 rad/s.

Use the following equations to determine the total acceleration at the outer radius = 32 in = 32/12 = 2.667 ft:

$$a_r = 2.667(6.3)^2 = 5105.84 \, \text{ft/s}^2$$
$$a_t = 2.667(0.157) = 0.418 \, \text{ft/s}^2$$

Total acceleration = $\sqrt{a_r^2 + a_t^2} = \sqrt{105.84^2 + 0.418^2} = 105.84 \, \text{ft/s}^2$

5. SolidWorks is used to determine the mass moment of inertia of the flywheel (Figure 11.6):

$$I_{zz} = 453088.10 \, \text{lb} - \text{in}^2 = \frac{453088.10}{32.2(12^2)} = 97.715 \, \text{lb} - \text{ft} - /\text{s}^2.$$

$$KE = \frac{1}{2}I_{zz}\omega^2 = \frac{1}{2}(97.715)(6.3)^2 = 1939.15 \, \text{ft} - \text{lb}.$$

Problem 11.3

A flywheel with a 58 in outside diameter, an inside diameter of 46 in, and weighing 400 lb initially rotates at 70 rpm clockwise. The flywheel decelerates from 70 rpm to a complete stop while completing 20 revolutions. The flywheel then remains stationary (i.e. 0 rpm) for 4 minutes. It then accelerates in a counter-clockwise direction to 88 rpm in 25 revolutions. Complete the following:

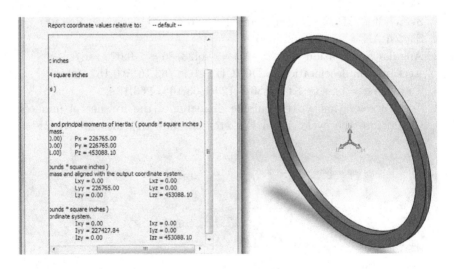

Fig. 11.6 Mass properties.

1. Sketch the complete cycle clearly showing critical points (clockwise rotation must be taken as positive 'y').
2. Calculate the angular acceleration and deceleration.
3. Find the total time to complete the cycle.
4. Find the magnitude of the total acceleration at the outside diameter of the flywheel 5 minutes into the cycle.
5. Find the kinetic energy 5 minutes into the cycle.

SolidWorks solution

$$\omega_o = \frac{2\pi(70)}{60} = 7.33 \qquad \omega = \frac{2\pi(88)}{60} = 9.215$$

$$\frac{1}{2}(t_1)(7.33) = 40\,\text{rev} = 20(2\pi) = 125.66\,\text{rad}$$

$$\therefore \ t_1 = 2(125.66)/7.33 = 34.28\,\text{s}$$

$$\frac{1}{2}(t_3)(9.215) = 25\,\text{rev} = 25(2\pi) = 157.079\,\text{rad}$$

$$\therefore \ t_3 = 2(157.079)/9.215 = 34.1\,\text{s}$$

1. Sketch the velocity–time diagram using SolidWorks (see Figure 11.7); this involves the following steps:
   Sketch line FA = 7.33 (initial angular velocity).
   Sketch AB = 34.28 (t_1).
   Sketch line BC = 240 (t_2 = 4 minutes).
   Sketch CD = 34.1 (t_3).
   Sketch line GD = 9.215 (final angular velocity).
   Sketch AE = 300 (5 minutes).
2. Angular acceleration = $-AF/AB = -7.33/34.28 = -0.214\,\text{rad/s}^2$.
   Acceleration deceleration = $DG/CD = 9.215/34.11 = 0.27\,\text{rad/s}^2$.
3. Total time = $t_1 + t_2 + t_3 = 34.26 + 240 + 34.11 = 308.4\,\text{s}$.

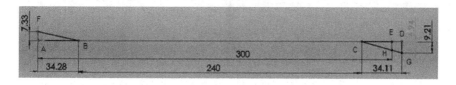

Fig. 11.7 Velocity–time diagram (complete cycle).

4. Total acceleration at the outside diameter of the flywheel 5 minutes (300 s) into the cycle:

   From E sketch the line HE; measure it. HE = 6.94 rad/s.

   Use the following equations to determine the total acceleration at the outer radius = 29 in = 32/12 = 2.417 ft:

   $$a_r = 2.417(6.94)^2 = 116.4\,\text{ft/s}^2$$
   $$a_t = 2.417(0.27) = 0.652\,\text{ft/s}^2$$

   Total acceleration = $\sqrt{a_r^2 + a_t^2} = \sqrt{116.4^2 + 0.652^2} = 116.41\,\text{ft/s}^2$

5. SolidWorks is used to determine the mass moment of inertia of the flywheel (Figure 11.8):

   Density = 0.28 lb/in^3.

   Mass = 401.31 lb (note that this is lb-mass, so we must divide it by 32.2).

   Volume = 1411.45 in^3.

   Surface area = 2430.84 in^2.

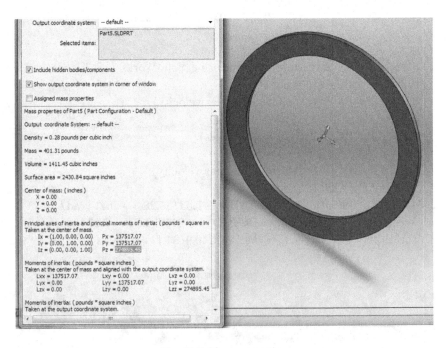

Fig. 11.8 Mass properties.

Center of mass (inches): X = 0.00; Y = 0.00; Z = 0.00.
Principal moments of inertia, taken at the center of mass:

$$Px = 137517.07$$
$$Py = 137517.07$$
$$Pz = 274895.45$$

Moment of inertia $= 274895.45/(32.2 \times 12 \times 12) = 59.285$ lb-ft-s^2.
This values agrees with theory using $1/2\,m(r_i^2 + r_o^2) = 1/2\frac{400}{32.2}(2.416^2 + 1.916^2) = 59.05$

$$KE = \frac{1}{2}I_{zz}\omega^2 = \frac{1}{2}(59.05)(6.94)^2 = 1422\,\text{ft} - \text{lb}$$

Problem 11.4

A pulley turns initially at 30 rpm clockwise then reverses its direction of rotation to 40 rpm counter-clockwise in 5 seconds, at a constant deceleration and acceleration rate. Determine:

1. The deceleration rate.
2. The total number of revolutions of the pulley during the 5-second interval.
3. The angular displacement of the pulley at t = 5 seconds.

SolidWorks solution

$$\omega_o = \frac{2\pi(30)}{60} = 3.14 \qquad \omega = \frac{2\pi(40)}{60} = 4.19$$

Applying the equation: $\omega = \omega_o + \alpha t$; $-4.19 = 3.14 + \alpha(5)$.
Solving for the angular acceleration leads to: $\alpha = (-4.19 - 3.14)/(5) = -1.47\,\text{rad/s}^2$.
Using these parameters, the line AB is sketched from $(0, 3.14)$ to $(5, -4.19)$ as shown in Figure 11.9. SolidWorks automatically gives the values of $t_1 = 2.14\,s$; $t_2 = 2.86$ s.
The total angle covered $= \theta = \frac{1}{2}(3.14)(2.14) + \frac{1}{2}(2.86)(4.19) = 9.35\,\text{rad} = \frac{9.35}{2\pi} = 1.49\,\text{rev}$.
The angular displacement $= \theta = -\frac{1}{2}(3.14)(2.14) + \frac{1}{2}(2.86)(4.19) = 2.63\,\text{rad} = \frac{2.63}{2\pi} = 0.42\,\text{rev}$.

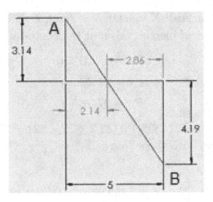

Fig. 11.9 Velocity–time diagram.

Problem 11.5

A rotating clockwise at 60 rpm clockwise has a torque applied to it that decelerates it to a stop, then accelerating it to 40 rpm counter-clockwise in 20 seconds. If the deceleration and acceleration are equal and constant, determine:

1. The angular displacement.
2. The revolution in each direction.

SolidWorks solution

$$\omega_o = \frac{2\pi(60)}{60} = 6.28 \qquad \omega = \frac{2\pi(40)}{60} = 4.19$$

Applying the equation: $\omega = \omega_o + \alpha t$; $-4.19 = 6.28 + \alpha(20)$.
Solving for the angular acceleration leads to: $\alpha = (-4.19 - 6.28)/(20) = -0.523\,\mathrm{rad/s^2}$.
Using these parameters, the line AB is sketched from (0, 6.28) to (20, −4.19) as shown in Figure 11.10. SolidWorks automatically gives the values of $t_1 = 12\,s$; $t_2 = 8\,s$.
The angular displacement $= \theta = \frac{1}{2}(12)(6.28) - \frac{1}{2}(8)(4.19) = 20.92\,\mathrm{rad}\ ccw$.
The angular displacement $= \theta = \frac{1}{2}(12)(6.28) = 37.68\,\mathrm{rad}\ cw$.
The angular displacement $= \theta = \frac{1}{2}(8)(4.19) = 16.76\,\mathrm{rad}\ ccw$.

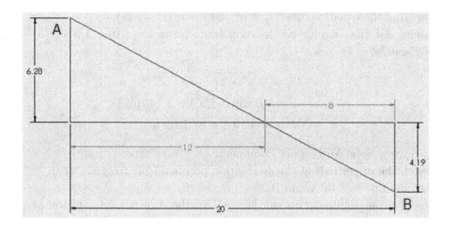

Fig. 11.10 Velocity–time diagram.

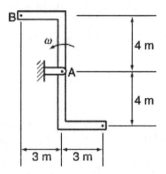

Fig. 11.11 A straight edge impeller model.

Problem 11.6

The object shown in Figure 11.11 rotates about A and accelerates from an initial speed of 10 rpm to 40 rpm in 5 seconds. At t = 5 seconds, determine the total acceleration of point B for the position shown.

SolidWorks solution

$$\omega_o = \frac{2\pi(10)}{60} = 1.047 \qquad \omega = \frac{2\pi(40)}{60} = 4.19$$

Applying the equation: $\omega = \omega_o + \alpha t$; $4.19 = 1.047 + \alpha(5)$

Solving for the angular acceleration leads to: $\alpha = (4.19 - 1.047)/(5) = 0.629 \, \text{rad/s}^2$

$$r = AB = \sqrt{3^2 + 4^2} = 5$$
$$a_t = r\alpha = 5(0.629 \, \text{rad/s}^2) = 3.145 \, \text{m/s}$$
$$a_r = r\omega^2 = 5(4.19^2) = 87.78 \, \text{m/s}$$

Open New SolidWorks Part Document.

Sketch the upper half of the description problem (due to symmetry).

Sketch a line to join A and B.

Sketch a perpendicular to the line AB in the direction of rotation at B (BC).

Sketch a line parallel to AB with a length of 87.78; this represents a_r (see Figure 11.12).

Sketch a line at the end of a_r that is perpendicular to it with a length of 3.14; this represents a_t (see Figure 11.12).

Sketch a line to close the triangle; this is the resultant, a_{tot} (total acceleration) (see Figure 11.12).

Problem 11.7

Point A has a total acceleration of $240 \, \text{in/s}^2 \downarrow \overline{35^\circ}$ (Figure 11.13). For member ABC, rotating clockwise, determine (a) angular velocity ω and (b) angular acceleration α.

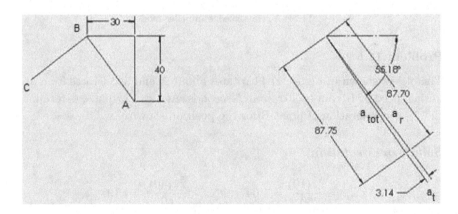

Fig. 11.12 Acceleration diagram.

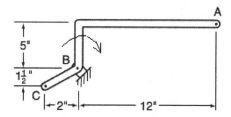

Fig. 11.13 Poblem description.

SolidWorks solution

Open New SolidWorks Part Document.

Choose a SCALE of 1:10.

Sketch the description problem.

Sketch a line to join A and B.

Sketch a perpendicular to the line AB in the direction of rotation at A (AD). The centripetal acceleration makes an angle $\theta = \tan^{-1}(50/120) = 22.62°$ with the horizontal (use the problem description to determine this angle) (see Figure 11.14).

Sketch a line BM with a length of 240; this represents a_{tot} (total acceleration) at an angle of 35° to the horizontal (given) (see Figure 11.14).

Sketch a line BN at an angle of 12.38° (35–22.62); this represents a_r (see Figure 11.14).

Sketch a line from M, perpendicular to MN; this represents a_t (see Figure 11.14).

Trim extended lines at vertex N.

Measure BN; this represents a_r.

Measure MN; this represents a_t.

The values are 234.42 and 51.46 respectively (see Figure 11.14).

Use the following equations to determine the parameters required:

$$r = AB = \sqrt{5^2 + 12^2} = 13$$

$$a_r = 23.442 = r\omega^2 = 5(\omega^2)$$

$$\therefore \ \omega = \sqrt{(23.442)/5} = 2.165 \,\text{rad/s}$$

$$a_t = 5.146 = r\alpha = 5\alpha$$

$$\therefore \ \alpha = 5.146/13 = 0.395 \,\text{rad/s}^2$$

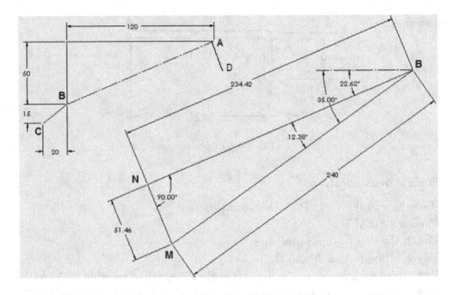

Fig. 11.14 Acceleration diagram.

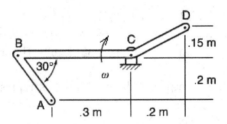

Fig. 11.15 Problem description.

Problem 11.8

The total acceleration of point A is 4 m/s ↑45° (Figure 11.15). For member ABCD determine (a) angular velocity ω and (b) angular acceleration α.

SolidWorks solution

Open New SolidWorks Part Document.
Choose a SCALE of 1:1.
Sketch the description problem.
Sketch a line to join A and C.
Sketch a perpendicular to the line AE in the direction of rotation at A (AE).

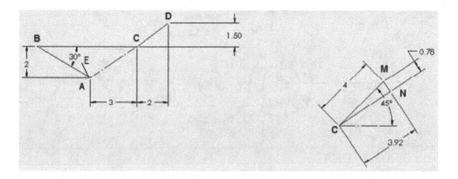

Fig. 11.16 Acceleration diagram.

The centripetal acceleration makes an angle $\theta = \tan^{-1}(0.2/0.3) = 33.7°$ with the horizontal (use the problem description to determine this angle) (see Figure 11.16).

Sketch a line CM with a length of 4; this represents a_{tot} (total acceleration) at an angle of 45° to the horizontal (given) (see Figure 11.16).

Sketch a line CN at an angle of 11.3° (45–33.7); this represents a_r (see Figure 11.16).

Sketch a line from M, perpendicular to MN; this represents a_t (see Figure 11.16).

Trim extended lines at vertex N.

Measure CN; this represents a_r.

Measure MN; this represents a_t.

The values are 3.92 and 0.78 respectively (see Figure 11.16).

Use the following equations to determine the parameters required:

$$r = AC = \sqrt{0.2^2 + 0.3^2} = 0.36$$

$$a_r = 3.92 = r\omega^2 = 0.36(\omega^2)$$

$$\therefore \ \omega = \sqrt{(3.92)/0.36} = 3.316 \, \text{rad/s}$$

$$a_t = 0.76 = r\alpha = 0.36\,\alpha$$

$$\alpha = 0.76/0.36 = 2.1 \ \text{rad/s}^2$$

Problem 11.9

In Figure 11.17 block B has a velocity of 12 in/s to the left and an acceleration of 2 in/s^2. Determine the following:

1. Velocity and acceleration of block A.
2. Total acceleration of point labeled '+' at this instant.

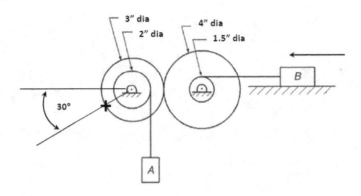

Fig. 11.17 Problem description.

SolidWorks solution

Open New SolidWorks Part Document
Velocity of Block A (see Figure 7.18).
Choose a SCALE of 1:10

$$v = \omega\, r; \quad \omega = \frac{v}{r} = const.$$

In the fist stage,

$$\omega = \frac{v_B}{r_B} = \frac{BE}{AE} = \frac{120}{75}$$

When AD = 200, the velocity is 320.
In the second stage (to the left),

$$\omega = \frac{v_2}{r_2} = \frac{HJ}{FJ} = \frac{320}{150}$$

When FK = 100, the velocity is 213.333. Therefore, the velocity of block A is:

$$v_A = 21.333\, in/s \downarrow$$

Acceleration of Block A (see Figure 11.19).
Choose a SCALE of 1:100

$$a = \alpha\, r; \quad \alpha = \frac{a}{r} = const.$$

In the first stage,

$$\alpha = \frac{a_B}{r_B} = \frac{BE}{AE} = \frac{200}{75}$$

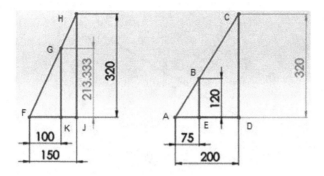

Fig. 11.18 Velocity relations.

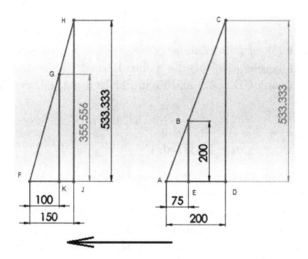

Fig. 11.19 Acceleration relations.

When AD = 200, the acceleration is 533.33.

In the second stage (to the left),

$$\alpha = \frac{a_2}{r_2} = \frac{HJ}{FJ} = \frac{533.33}{150}$$

When FJ = 100, the acceleration is 355.556. Therefore, the acceleration of block A is:

$$a_A = 3.555\, in/s^2 \downarrow$$

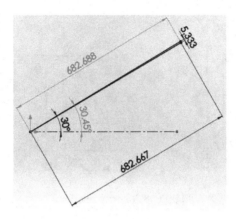

Fig. 11.20 Total acceleration.

Total Acceleration of Point Labeled '+'

The total acceleration of the point labeled '+' is obtained from Figure 11.18 (using CD = 32) and Figure 11.19 using (CD = 5.333):

$$a_r = \frac{v_2^2}{r_2} = \frac{32^2}{1.5} = 682.667 \,\text{in/s}^2$$

$$a_t = 5.333 \,\text{rad/s}^2$$

Sketch a line at an angle of 30° to the horizontal, with a length of 682.667; this represents a_r (see Figure 11.20).

Sketch a line at the end of a_r that is perpendicular to it with a length of 5.333; this represents a_t (see Figure 11.20).

Sketch a line to close the triangle; this is the resultant, a_{tot} (total acceleration) (see Figure 11.20).

Summary

This chapter has described using some examples how SolidWork is used to solve angular kinematics problems in applied mechanics. It can be observed that SolidWorks offers very simple approach to solving.

Exercises

P1. Block A is dropping at a velocity of 600 mm/s and an acceleration of 100 mm/s² in Figure P1. Determine the linear velocity and acceleration of block B. Also, determine the tangential velocity and acceleration at the point of contact between the two cylinders.

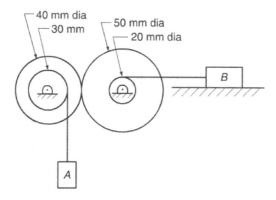

Fig. P1

P2. A flywheel having a 64 in outside diameter, an inside diameter of 56 in, and weighing 500 lb is initially rotating at 60 rpm clockwise. The flywheel decelerates from 60 rpm to a complete stop while completing 40 revolutions. The flywheel then remains stationary (i.e. 0 rpm) for 4 minutes. It then accelerates in a counter-clockwise direction to 95 rpm in 50 revolutions. Complete the following:

1. Sketch the complete cycle clearly showing critical points (clockwise rotation must be taken as positive 'y'). (Compare your solution with Figure P2.)
2. Calculate the angular acceleration and deceleration.
3. Find the total time to complete the cycle.
4. Caluculate the magnitude of the total acceleration at the outside diameter of the flywheel 4 minutes into the cycle.
5. Find the kinetic energy 4 minutes into the cycle.

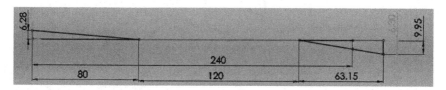

Fig. P2 Velocity–time diagram (complete cycle).

Reference

Walker, K. M., *Applied Mechanics for Engineering Technologists*, 8th Edition, Prentice Hall, Upper Saddle River, NJ, 2007.

Chapter 12

Kinematics: Plane Motion of Mechanisms I

When you complete this chapter you will:

- Have understood the Applied Mechanics Virtual Library (AMVL) concept using SolidWorks.
- Have understood the concept of using SolidWorks to draw velocity and acceleration diagrams.

Applied Mechanics Virtual Library (AMVL) Concept

In applied mechanics, students should attempt to utilize *logic, experience,* and *visualization* as much as possible. One of the main challenges in solving engineering mechanics problems is the ability of students to visualize how machine members that constitute an assembly move in space. Experienced designers may be able to visualize correctly how machine members move, but this can be a considerable challenge to students who have limited industrial experience.

AMVL concept (Figure 12.1) was devised by the author using Solid-Works to *model* mechanisms in a machine which are animated so that students can understand how machine parts move while in operation. The other components of the AMVL initiative include the actual *geometric analysis for motion* (kinematics) and the effect of force (kinetics) on motion which are handled using SolidWorks. When AMVL was used to solve dynamics problems, students could visualize how the mechanisms moved, and they unanimously affirmed that AMLV is an extremely useful tool for solving dynamics problems. Consequently, the AMVL project facilitated the teaching and learning of applied mechanics very significantly and the measured outcome was that students had much better understanding of the course.

The basis for the success of AMLV is simply the fact that SolidWorks is user-friendly, efficient, and an effective CAD software for modeling (which

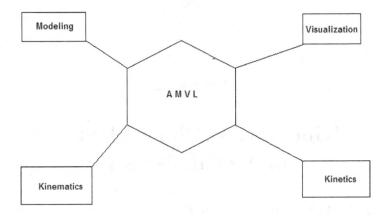

Fig. 12.1 AVML concept for teaching applied mechanics.

supports visualization) and analysis of machine members. This aspect of the features of SolidWorks seems not to have been explored by many users of SolidWorks.

This chapter is divided into two parts. Part I of this chapter concentrates on the modeling of mechanisms for plane motion using the SolidWorks Motion Analysis tool. This aspect covers the modeling and visualization components of the AMVL framework. Part II of this chapter concentrates on the analysis of mechanisms for plane motion using SolidWorks. This aspect covers the kinematics and kinetics components of the AMVL framework.

Modeling of Mechanisms: Kinematics — Plane Motion using SolidWorks

Case 1: Problem Description

The angular velocity of AB in Figure 12.2 is 400 rpm clockwise. Model the mechanism and animate motion.

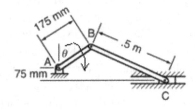

Fig. 12.2 Problem description.

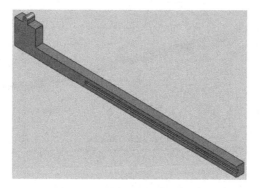

Fig. 12.3 Body.

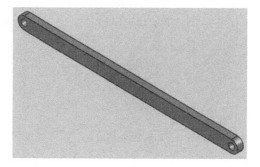

Fig. 12.4 Link1.

SolidWorks modeling solution

SolidWorks is used to model the components of the mechanism of Figure 12.2. The components of the mechanism are as follows:

1. Body (Figure 12.3).
2. Link1 (Figure 12.4).
3. Link2 (Figure 12.5).
4. Block (Figure 12.6).
5. Pin (Figure 12.7).

The assembly of the mechanism is shown in Figure 12.8.

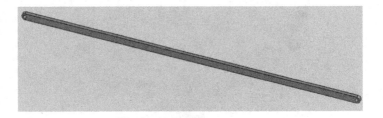

Fig. 12.5 Link2.

Fig. 12.6 Block.

Fig. 12.7 Pin.

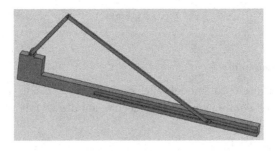

Fig. 12.8 Assembly of the mechanism.

SolidWorks motion analysis

The SolidWorks Motion Analysis file and the AVI file are available so that users can easily visualize the motion of the mechanism before analyzing the motion.

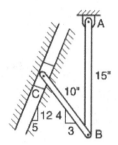

Fig. 12.9 Problem description.

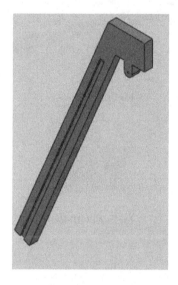

Fig. 12.10 Body.

Case 2: Problem Description

The velocity of slider C is 26 in/s downward (Figure 12.9). Model the mechanism and animate motion.

SolidWorks modeling solution

SolidWorks is used to model the components of the mechanism of Figure 12.9. The components of the mechanism are as follows:

1. Body (Figure 12.10).
2. Link1 (Figure 12.11).

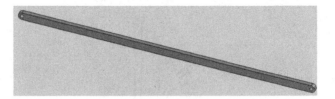

Fig. 12.11 Link1.

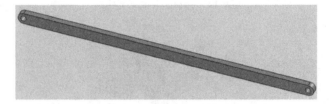

Fig. 12.12 Link2.

Fig. 12.13 Block.

Fig. 12.14 Pin.

3. Link2 (Figure 12.12).
4. Block (Figure 12.13).
5. Pin (Figure 12.14).

The assembly of the mechanism is shown in Figure 12.15.

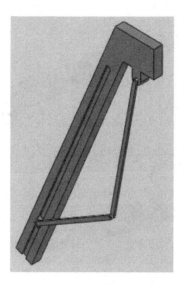

Fig. 12.15 Assembly of the mechanism.

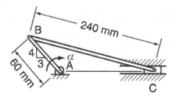

Fig. 12.16 Problem description.

Case 3: Problem Description

At the instant shown in Figure 12.16, AB has an angular acceleration of $8\,\text{rad/s}^2$ clockwise. Model the mechanism and animate motion.

SolidWorks modeling solution

SolidWorks is used to model the components of the mechanism of Figure 12.16. The components of the mechanism are as follows:

1. Body (Figure 12.17).
2. Link1 (Figure 12.18).

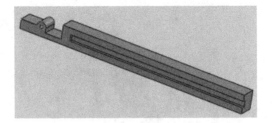

Fig. 12.17 Body.

Fig. 12.18 Link1.

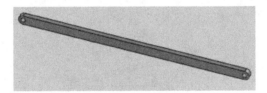

Fig. 12.19 Link2.

Fig. 12.20 Block.

3. Link2 (Figure 12.19).
4. Block (Figure 12.20).
5. Pin (Figure 12.21).

The assembly of the mechanism is shown in Figure 12.22.

Fig. 12.21 Pin.

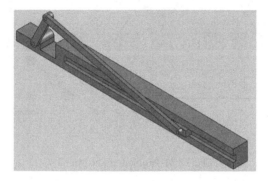

Fig. 12.22 Assembly of the mechanism.

SolidWorks motion analysis

The SolidWorks Motion Analysis file and the AVI file are available so that users can easily visualize the motion of the mechanism before analyzing the motion.

Case 4: Problem Description

The input arm O_1A rotates 10 rad/s clockwise in Figure 12.23. Model the mechanism and animate motion.

SolidWorks modeling solution

SolidWorks is used to model the components of the mechanism of Figure 12.23. The components of the mechanism are as follows:

1. Body (Figure 12.24).
2. Link1 (Figure 12.25).
3. Link2 (Figure 12.26).
4. Link3 (Figure 12.27).
5. Block (Figure 12.28).
6. Pin (Figure 12.29).

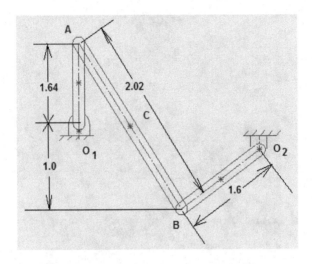

Fig. 12.23 Problem description.

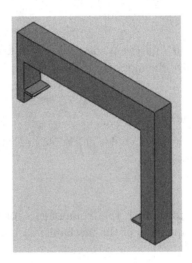

Fig. 12.24 Body.

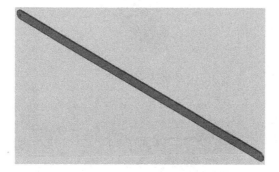

Fig. 12.25 Link1.

Fig. 12.26 Link2.

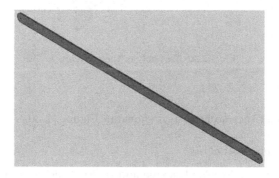

Fig. 12.27 Link3.

Fig. 12.28 Block.

Fig. 12.29 Pin.

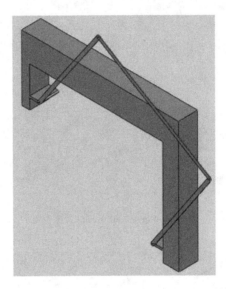

Fig. 12.30 Assembly of the mechanism.

The assembly of the mechanism is shown in Figure 12.30.

SolidWorks Motion Analysis

The SolidWorks Motion Analysis file and the AVI file are available so that users can easily visualize the motion of the mechanism before analyzing the motion.

Case 5: Problem Description

The input arm OA rotates 5 rad/s clockwise in Figure 12.31. Model the mechanism and animate motion.

SolidWorks modeling solution

SolidWorks is used to model the components of the mechanism of Figure 12.31. The components of the mechanism are as follows:

1. Body (Figure 12.32).
2. Link1 (Figure 12.33).
3. Link2 (Figure 12.34).
4. Link3 (Figure 12.35).
5. Block (Figure 12.36).
6. Pin (Figure 12.37).

The assembly of the mechanism is shown in Figure 12.37.

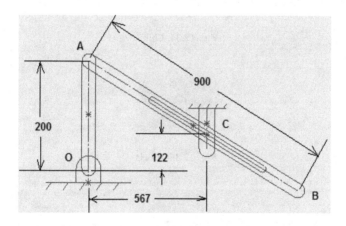

Fig. 12.31 Problem description.

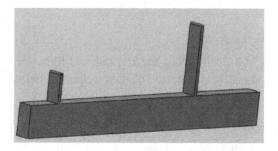

Fig. 12.32 Body.

Fig. 12.33 Link1.

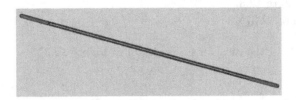

Fig. 12.34 Link2.

Fig. 12.35 Block.

Fig. 12.36 Pin.

SolidWorks motion analysis

The SolidWorks Motion Analysis file and the AVI file are available so that users can easily visualize the motion of the mechanism before analyzing the motion.

Case 6: Problem Description

Roller A of Figure 12.38 moves downward to the left at 45 in/s. Determine the linear velocity of point E.

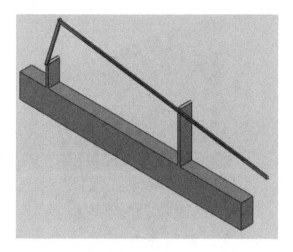

Fig. 12.37 Assembly of the mechanism.

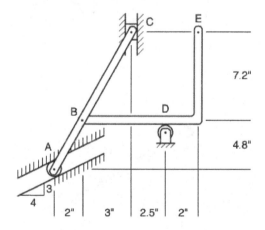

Fig. 12.38 Problem description.

SolidWorks modeling solution

SolidWorks is used to model the components of the mechanism of Figure 12.38. The components of the mechanism are as follows:

1. Body (Figure 12.39).
2. Link1 (Figure 12.40).
3. Link2 (Figure 12.41).

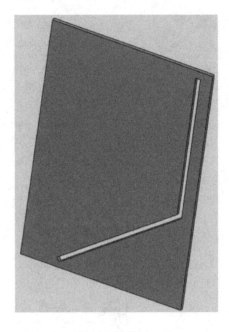

Fig. 12.39 Body.

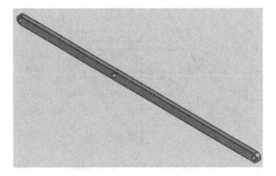

Fig. 12.40 Link1.

4. Block (Figure 12.42).
5. Pin (Figure 12.43).
6. Roller (Figure 12.44).

The assembly of the mechanism is shown in Figure 12.45.

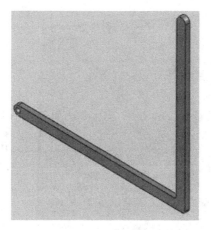

Fig. 12.41 Link2.

Fig. 12.42 Block.

Fig. 12.43 Pin.

Fig. 12.44 Roller.

SolidWorks motion analysis

The SolidWorks Motion Analysis file and the AVI file are available so that users can easily visualize the motion of the mechanism before analyzing the motion.

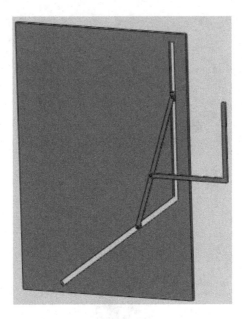

Fig. 12.45 Assembly of the mechanism.

Velocity Diagrams

This section involves the construction of diagrams which need to be done accurately and to a suitable scale. Traditionally, students and practitioners need a drawing board, ruler, compass, protractor, and triangles and should have some drawing skills. This section introduces SolidWorks as a useful CAD tool to replace the out-dated approach of drawing board and the tools listed above.

Absolute and relative velocity

An absolute velocity is the velocity of a point measured from a fixed point (normally the ground or anything rigidly attached to the ground and not moving). Relative velocity is the velocity of a point measured relative to another that may itself be moving.

Tangential velocity

Consider a link AB pinned at A and revolving about A at angular velocity ω (see Figure 12.46). Point B moves in a circle relative to point A but its velocity is always tangential and hence at 90° to the link. A convenient

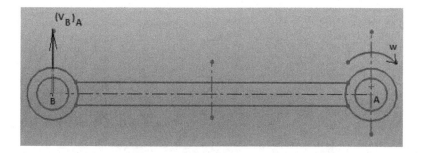

Fig. 12.46 Tangential velocity.

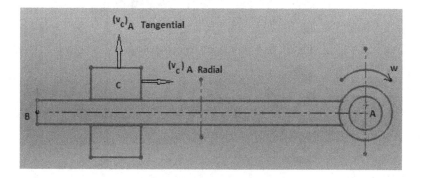

Fig. 12.47 Radial velocity.

method of denoting this tangential velocity is $(v_B)_A$ meaning the velocity of B relative to A. This method is not always suitable.

Radial velocity

Consider a sliding link C that can slide on link AB (see Figure 12.47). The direction can only be radial relative to point A as shown. If the link AB rotates about A at the same time then link C will have radial and tangential velocities. Note that both the radial and tangential velocities are denoted the same so the tags radial and tangential are added.

The sliding link has two relative velocities, the radial and tangential. They are normal to each other and the true velocity relative to A is the vector sum of both added as shown in Figure 12.48. Not that lower case letters are used on the vector diagrams. The two vectors are denoted by c_1 and c_2. The velocity of link C relative to point is the vector ac_2.

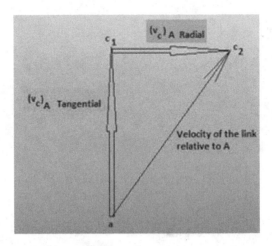

Fig. 12.48 Radial and tangential velocity.

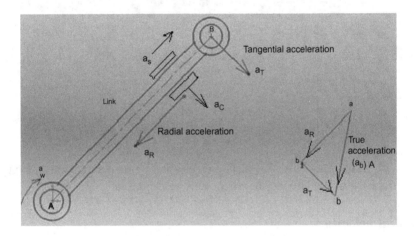

Fig. 12.49 Acceleration diagram.

Acceleration Diagrams

In the design of machine elements, it is important to determine the acceleration of links since acceleration produces inertia forces in the link which stress the component parts of the mechanism. Accelerations may be relative or absolute in the same way as described in velocity. We will present methodology for producing acceleration diagrams (see Figure 12.49) and consider four forms of accelerations, namely, tangential (a_T), radial (a_R), sliding (a_S), and Coriolis (a_C).

Tangential acceleration

Tangential acceleration only occurs if the link has angular acceleration $\alpha\,\text{rad/s}^2$. Consider a link AB with angular acceleration about A.

Point B will have both radial and tangential acceleration relative to point A. The true acceleration of point B relative to A is the vector sum of them. This will require an extra point. We will use b_1 and b on the vector diagram as shown. Point B is accelerating about a circular path and its direction is tangential (at right angles to the link). It is designated a_T and calculated using $a_T = \alpha \times AB$. The vector starts at b1 and ends at b. The choice of letters and notation are arbitrary but must be logical to aid and relate to the construction of the diagram.

Radial (centripetal) acceleration

A point rotating about a centre at radius R has a tangential velocity v and angular velocity ω and it is continually accelerating towards the center even though it never moves any closer. This is radial (centripetal) acceleration and it is caused by the constant change in direction. It follows that the end of any rotating link will have a centripetal acceleration towards the opposite end.

The relevant equations are: $v = \omega R$; $a = \omega^2 R$ or $a = v^2/R$. The construction of the vector for radial acceleration of B relative to A is in a radial direction so a suitable notation might be $\mathbf{a_R}$. It is calculated using the following equation:

$$a_R = \omega^2 R \quad \text{or} \quad a_R = v^2/R$$

Note that the direction is towards the center of rotation but the vector starts at a and ends at b_1. It is very important to get this the right way round otherwise the complete diagram will be wrong.

Sliding acceleration

A sleeve that slides on a rotating link will slide outward along the link.

Coriolis acceleration

Consider a link rotating with an angular velocity ω rad/s and accelerating at α rad/s^2. On the link is a sliding element moving away from the center of rotation with velocity $v_R = \frac{dR}{dt}$ (which is positive if getting larger). The link has a tangential velocity $v_T = \omega R$. It can be shown that the tangential acceleration is not simply as is the case for a constant radius but an extra term is added which is called the Coriolis acceleration which is a result of solving problems involving changing radius. The acceleration is therefore given as:

$$a_T = 2\,\omega\,v_R + \alpha\,R$$

The Coriolis acceleration is $2\omega v_R$. The steps for drawing acceleration diagram are as follows:

1. Start with the radial (centripetal) acceleration (always pointing INWARDS toward the point of rotation along the rotating link).
2. At the end of the radial (centripetal) acceleration add the tangential acceleration (if it is given) perpendicular to the rotating link.
3. Add sliding (along the link) and Coriolis (perpendicular to the link) accelerations if they exist.

Problem 12.1

Bar AB rotates at 5 rad/s clockwise and accelerates at 2 rad/s^2 clockwise. The dimensions are $L_1 = 20\,\text{cm}$, $L_2 = 50\,\text{cm}$, $L_3 = 60\,\text{cm}$, and $y = 10\,\text{cm}$. Determine the linear acceleration of C at its position in Figure 12.50.

SolidWorks modeling solution

Parts and assembly modeling for plane motion kinematics

Body: Sketch shown in Figure 12.51. Extrude 1 cm.

There are two holes 2.5 cm diameter, 100 cm apart horizontally and 10 cm apart vertically (see Figure 12.52).

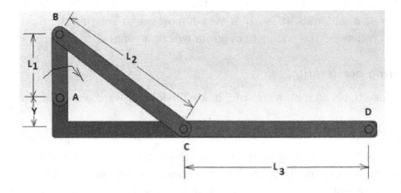

Fig. 12.50 Problem description.

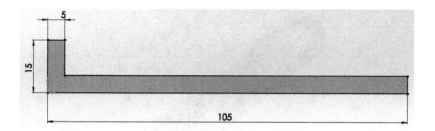

Fig. 12.51 Sketch for body.

Fig. 12.52 Body.

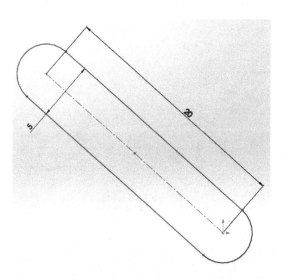

Fig. 12.53 Link1.

Link1: Sketch shown in Figure 12.53. Extrude 1 cm.
There are two holes, 2.5 cm diameter, 20 cm apart (see Figure 12.54).
Link2: Sketch shown in Figure 12.55. Extrude 1 cm.
There are two holes, 2.5 cm diameter, 50 cm apart (see Figure 12.56).

Fig. 12.54 Link2.

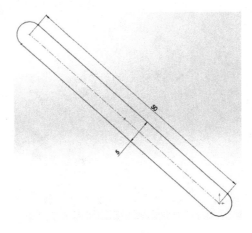

Fig. 12.55 Link2.

Link3: Extrude 1 cm.

There are two hole 2.5 cm diameter, 60 cm apart (see Figure 12.57).

Pin1: Circle 2.5 cm diameter (see Figure 12.58).

Extrude 2 cm.

Pin2: Circle 2.5 cm diameter (see Figure 12.59).

Extrude 6 cm.

Assembly

The assembly of the mechanism is shown in Figure 12.60.

Fig. 12.56 Link2.

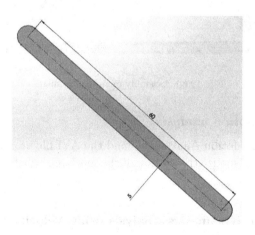

Fig. 12.57 Link3.

Fig. 12.58 Pin.

Fig. 12.59 Pin.

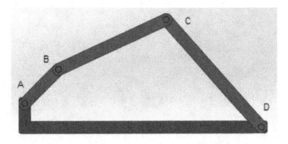

Fig. 12.60 Assembly of the mechanism.

SolidWorks motion analysis

The SolidWorks Motion Analysis file and the AVI file are available so that users can easily visualize the motion of the mechanism before analyzing the motion.

Plane Motion: Kinematics Analysis using Velocity Diagram

Figure 12.61 shows the kinematic analysis solution using velocity and acceleration diagrams; we will come to this method later in the chapter.

Problem 12.2

Referring to Figure 12.62, the velocity of link C/B is 18 m/s downwards to the left. Determine:

1. Angular velocity of link AB.
2. Velocity of piston C.

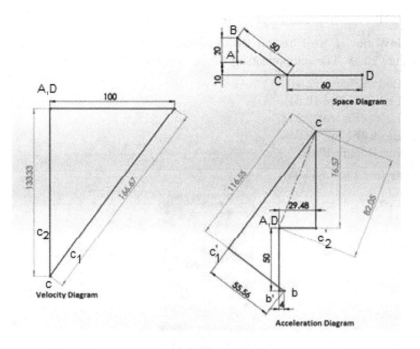

Fig. 12.61 Kinematic analysis solution using velocity diagram.

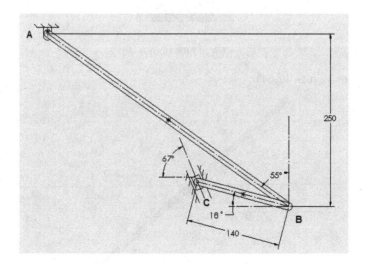

Fig. 12.62 Mechanical linkage (all dimensions in mm).

SolidWorks modeling solution

SolidWorks is used to model the components of the mechanism of Figure 12.58. The components of the mechanism are as follows:

1. Body (Figure 12.63).
2. Link1 (Figure 12.64).
3. Link2 (Figure 12.65).
4. Block (Figure 12.66).
5. Pin (Figure 12.67).

Body (see Figure 12.63).

Fig. 12.63 Body.

Link1 (see Figure 12.64).

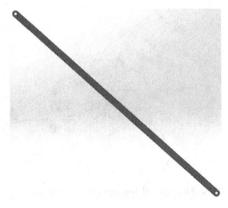

Fig. 12.64 Link1.

Link2 (see Figure 12.65).

Fig. 12.65 Link2.

Block (see Figure 12.66).

Fig. 12.66 Block.

Pin (see Figure 12.67).

Fig. 12.67 Pin.

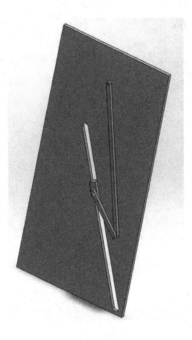

Fig. 12.68 Assembly of mechanism.

Assembly

The assembly of the mechanism is shown in Figure 12.68.

SolidWorks motion analysis

The SolidWorks Motion Analysis file and the AVI file for Figure 12.69 are available so that users can easily visualize the motion of the mechanism before analyzing the motion.

Plane Motion: Kinematics Analysis using Velocity Diagram

The solution to the kinematic analysis for this problem is shown in Figure 12.70.

The body of the machine is A,Q. The block C slides in the slot Q at the left.

Choose scale: 10:1.

The velocity of C with respect to B is bc = 18 m/s which starts at b and is perpendicular to BC downward to the left.

The velocity of B with respect to A is perpendicular to AB through B.

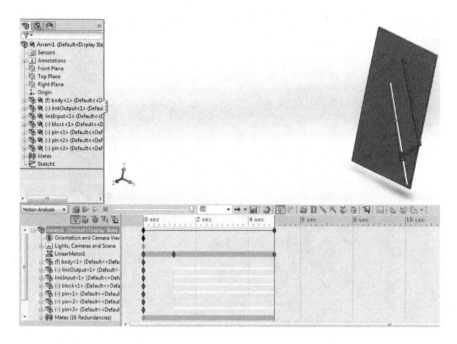

Fig. 12.69 SolidWorks Motion Analysis file.

C slides along the slot at Q; from C draw a line parallel to the slot. The intersection of two previous lines, fix A,Q.

From the velocity diagram, $V_b = 13.92 \, \text{m/s}$.

Therefore, the angular velocity of link AB $= V_b/|AB| = \frac{13.92}{(.25/\cos 55°)} = \frac{13.92}{0.4358} = 31.96 \, \text{rad/s}$

$$V_c = A, Q - c = 6.2 \text{ m/s}$$

Plain Motion Analysis using SolidWorks: Tutorials

Problem 12.3

Starting from the same point, car A travels north at $40 \, \text{km/h}$ and car B travels east at $60 \, \text{km/h}$ (Figure 12.71). Determine the velocity of B with respect to A.

SolidWorks solution for velocity diagram

The SolidWorks solution based on velocity diagram (VD) is shown in Figure 12.72.

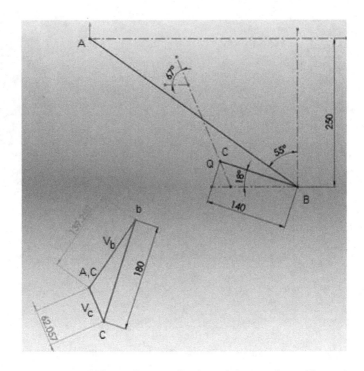

Fig. 12.70 Velocity diagram showing solution to the problem.

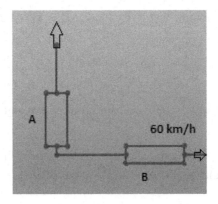

Fig. 12.71 Problem definition.

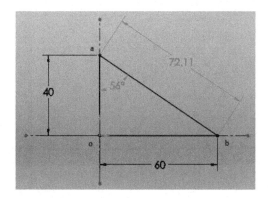

Fig. 12.72 Velocity diagram.

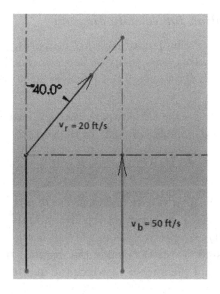

Fig. 12.73 Problem definition.

Problem 12.4

A footballer receiver runs straight downfield (north), turns 40° to his right and, maintaining his speed at 20 ft/s, catches a football that is traveling at 50 ft/s due north (see Figure 12.73). Determine the velocity of the ball with respect to the receiver.

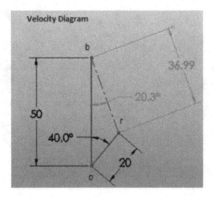

Fig. 12.74 Velocity diagram.

SolidWorks solution for velocity diagram

1. Choose o as the starting point (arbitrary in space).
2. $ob = 50\,ft./s \uparrow$ (o–b is the velocity of Receiver (north); given).
3. $or = 20\,ft./s$ (o–b is the velocity of Receiver at 40° to north; given).

From Figure 12.74, b–r is the velocity of the ball with respect to the receiver; the value is 37 ft/s at an angle of 20.3° to the vertical.

Problem 12.5

Suppose now that the quarterback and receiver are located (distances are in yards) as in Figure 12.75. As before, the receiver and the ball have velocities of 20 ft/s and 50 ft/s, respectively; and the ball is released the instant the receiver makes his 40° turn. Determine the distance traveled by the ball before it is caught and the angle θ at which the quarterback must lead his receiver.

SolidWorks solution for velocity diagram

Sketch the geometry of motion (Figure 12.75).

 Measure the angle (23°) and distance (30.09).

 Therefore, the ball travels 30 yards and the quarterback leads the receiver by 23°.

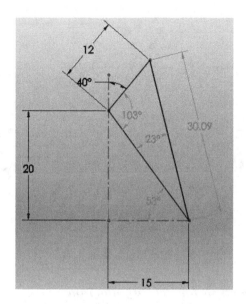

Fig. 12.75 Velocity diagram.

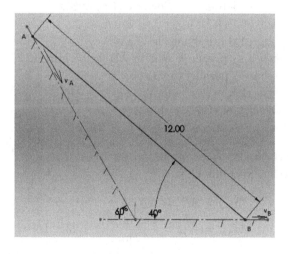

Fig. 12.76 Problem definition.

Problem 12.6

Determine (a) the linear velocity of end A of the bar shown in Figure 12.76 if the velocity of B is 16 in/s to the right and (b) the angular velocity of AB.

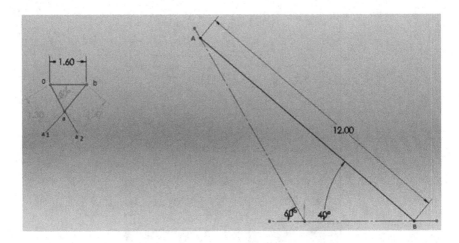

Fig. 12.77 Velocity diagram.

SolidWorks solution for velocity diagram

1. Choose o as the starting point (arbitrary in space).
2. $ob = 16\,in./s \rightarrow$ (o–b is the velocity of B; given).
3. Choose a scale of 1:10.
4. Therefore, ob = 1.6 (applying the scale).
5. From b sketch b-a$_1$ perpendicular to AB (see Figure 12.77).
6. From o sketch o-a$_2$ parallel to v_A.
7. a is the intersection of b-a$_1$ and o-a$_2$.
8. o–a is the velocity of end A = 1.30.

a. Applying the scale, o–a = 13 in/s at 60° to horizontal.
b Angular velocity of link CD,

$$v_{AB} = AB \times \omega_{AB}$$
$$\omega_{AB} = \frac{ab}{AB} = \frac{14.7\,in./s}{12\,in.} = 1.23\,rad/s$$

Problem 12.7

Bar AB of the linkage shown in Figure 12.78 rotates at 20 rad/s clockwise. For the position shown, determine the velocity of pin C and the angular velocity of link CD.

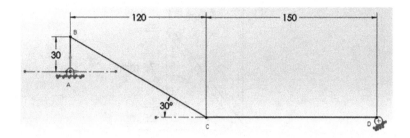

Fig. 12.78 Problem definition.

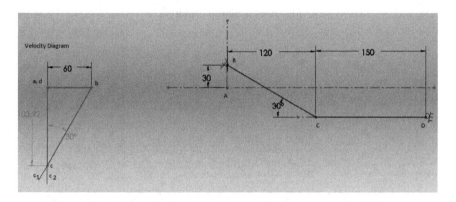

Fig. 12.79 The velocity diagram is on the left-hand side of the figure.

SolidWorks solution for velocity diagram

1. Choose a,d as the starting point (arbitrary in space).
2. $ab = r\omega = (30)(20\,\text{rad/s}) = 600\,\text{mm/s} \rightarrow$ (a–b is the velocity of AB; perpendicular to AB).
3. Choose a scale of 10:1.
4. Therefore, ab = 60 (applying the scale).
5. From b sketch b–c_1 perpendicular to BC (see Figure 12.79).
6. From a,d sketch d–c_2 perpendicular to BC.
7. c is the intersection of b–c_1 and d–c_2.
8. d–c_1 is the velocity of pin C = 103.92.
9. Applying the scale, d – c_1 = 1039.2 mm/s downward.
10. Angular velocity of link CD,

$$v_c = r\omega_{CD}$$
$$\omega_{CD} = \frac{1039.2\,\text{mm/s}}{150\,\text{mm}} = 6.93\,\text{rad/s}$$

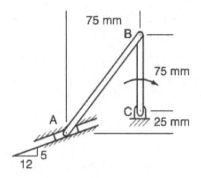

Fig. 12.80 Problem definition.

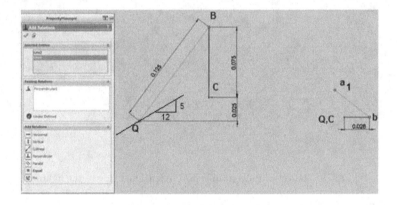

Fig. 12.81 Velocity of point B relative to C.

Problem 12.8

Bar BC in Figure 12.80 has an angular velocity of 3.73 rad/s clockwise. Determine (a) the angular velocity of AB and (b) the velocity of the piston A.

SolidWorks solution for velocity diagram

1. Choose Q,C as the starting point (block A slides in Q) (arbitrary in space) (see Figure 12.81).
2. $Cb = r\omega = (0.075)(3.75\,\text{rad/s}) = 0.28\,\text{mm/s} \rightarrow$ (C–b is velocity of CB; perpendicular to CB).
3. Choose a scale of 10:1.
4. Therefore, C–b = 0.028 (applying the scale).

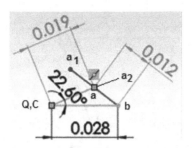

Fig. 12.82 Velocity of point A relative to the slot.

5. From b sketch b–a_1 perpendicular to QB (see Figure 12.81).
6. From Q,C sketch Q–a_2 at an angle of $\theta = \tan^{-1}(5/12) = 22.6°$ to the horizontal (along the slot) (see Figure 12.82).
7. a is the intersection of b–a_1 and Q–a_2.
8. a–b is the velocity of AB $= 0.12$ (taking the scale into consideration).
9. Q–a is the velocity of piston A $= 0.19$ (taking the scale into consideration).

The length of AB $= \sqrt{0.1^2 + 0.075^2} = 0.125\,\text{m}$.

(a) Angular velocity of AB is:

$$v_{AB} = AB \times \omega_{AB}$$
$$\omega_{AB} = \frac{ab}{AB} = \frac{0.12\ \text{m/s}}{0.125\,\text{m}} = 0.96\,\text{rad/s}$$

(b) The velocity of the piston A $= 0.19$ m/s (already obtained).

Problem 12.9

The velocity of point A is 10 m/s to the right, for the system shown (Figure 12.83). Determine the angular velocity of AC and the linear velocity of point C.

SolidWorks solution

The sketch of the mechanism is first produced as shown in Figure 12.84.

The velocity diagram is drawn using SolidWorks Sketching Tools. The horizontal and inclined hatched planes are considered as belonging to the body of the mechanism or machine (grounded).

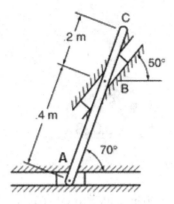

Fig. 12.83 Problem definition.

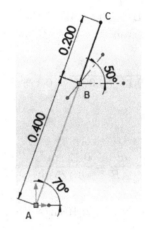

Fig. 12.84 Sketch of the mechanism.

1. Choose O,Q as the starting point (arbitrary in space) (see Figure 12.85).
2. Choose a scale of 1:10.
3. $Oa = 1\,m/s \rightarrow$ (O–a is velocity of A; given as being to the right).
4. Sketch a line a–b_1 perpendicular to AB (see Figure 12.85).
5. From O,Q sketch Q–b_2 at an angle of $\theta = 50°$ to the horizontal (along slot) (see Figure 12.86).
6. b is the intersection of a–b_1 and Q–b_2 (see Figure 12.87).

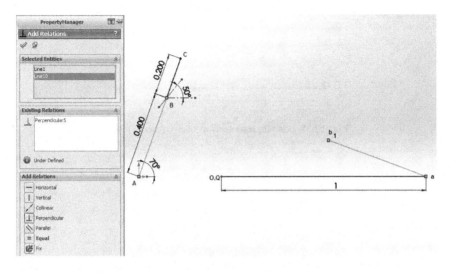

Fig. 12.85 Velocity of point A.

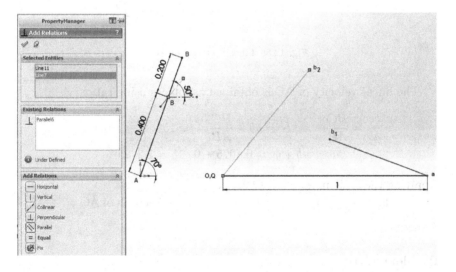

Fig. 12.86 Velocity of point B relative to the slot Q.

7. a–b is the velocity of AB = 8.15 (taking the scale into consideration).
8. Using ratio, **ac** is the linear velocity of AC = 12.225 (taking the scale into consideration) (see Figure 12.88).

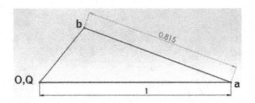

Fig. 12.87 b is the intersection of a–b₁ and Q–b₂.

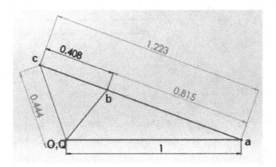

Fig. 12.88 Linear velocity of AC.

The linear velocity of AC is obtained as follows using ratio:

$$\frac{bc}{ab} = \frac{BC}{AB}; \quad \therefore \ bc = ab \times \frac{BC}{AB} = 0.815 \times \frac{0.2}{0.4} = 0.4075$$
$$ac = ab + bc = 0.4075 + 0.815 = 1.2225$$

Applying the scale used, ac = 12.225 m/s.

Angular velocity of AC = linear velocity of AC/length of AC = $\frac{12.225}{0.6}$ = 20.375 rad/s.

Problem 12.10

Bar AB rotates at 5 rad/s clockwise and accelerates at 2 rad/s² clockwise. Determine the linear acceleration of C at its position in Figure 12.89.

SolidWorks solution

The sketch of the mechanism is first produced as shown in Figure 12.90. The velocity diagram is drawn using SolidWorks Sketching Tools. Joints A

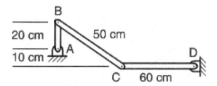

Fig. 12.89 Problem definition.

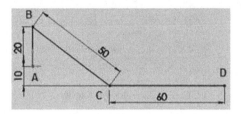

Fig. 12.90 Sketch of the mechanism.

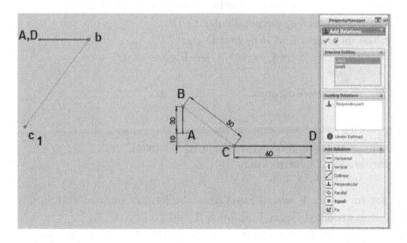

Fig. 12.91 Velocity of point C relative to point B.

and D are considered as belonging to the body of the mechanism or machine (grounded).

1. Choose A,D as the starting point (arbitrary in space) (see Figure 12.91).
2. $Ab = r\omega = (20)(5\,\text{rad/s}) = 100\,\text{cm/s} \rightarrow$ (A–b is velocity of AB; perpendicular to AB).

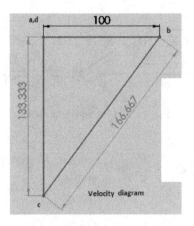

Fig. 12.92 Velocity diagram.

3. Choose a scale of 1:1.
4. From b sketch b–c_1 perpendicular to BC (see Figure 12.91).
5. From D sketch D–c_2 perpendicular to DC.
6. c is the intersection of b–c_1 and D–c_2.
7. D–c is the velocity of C = 133.3 (see Figure 12.92).

Rule for acceleration diagram:

1. Start with radial acceleration (obtained the velocity diagram).
2. Follow with tangential acceleration (if it exists).
3. Continue from joint to joint.

Rule for radial acceleration: Start from the joint of interest and go beyond where it is attached (start from joint B and go beyond A).

Rule for tangential acceleration: It is in the direction of the velocity.

Tangential acceleration of B with respect to joint A $= r\alpha =$ $(20)(2\,\mathrm{rad/s^2}) = 40\,\mathrm{rad/s^2} \rightarrow$.

Radial acceleration of B with respect to joint A $= r\omega^2 = \frac{ab^2}{AB} = \frac{100^2}{25} =$ $500\,\mathrm{rad/s^2} \downarrow$.

Radial acceleration of C with respect to joint B $= r\omega^2 = \frac{cb^2}{CB} = \frac{166.667^2}{50} =$ $555.557\,\mathrm{rad/s^2}$.

Radial acceleration of C with respect to joint D $= r\omega^2 = \frac{cd^2}{CD} = \frac{133.33^2}{60} =$ $296.29\,\mathrm{rad/s^2}$.

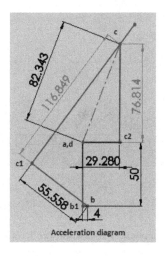

Fig. 12.93 Acceleration diagram.

	Radial (Direction as in **VD**)	Tangential
B w.r.t. A	$500\,\mathrm{rad/s^2}$ ↓	$40\,\mathrm{rad/s^2}$ →
C w.r.t. B	$555.557\,\mathrm{rad/s^2}$ ↓	Unknown
C w.r.t. D	$296.29\,\mathrm{rad/s^2}$ ↓	Unknown

1. Choose a,d as the starting point (arbitrary in space) (see Figure 12.93).
2. Choose a scale of 1:10.
3. $a\text{–}b_1 = 50$ (Radial acceleration: Start from joint A and go in the direction of joint B to A).
4. $b_1\text{–}b = 4$ (Tangential acceleration: Given; in the direction of velocity of B, to the right).
5. $b\text{–}c_1 = 55.558$ (Radial acceleration: Start from C and go in the direction of C to B).
6. $c_1\text{–}\ = $ Unknown (Tangential acceleration: In the direction of velocity of C w.r.t. B).
7. $d\text{–}c_2 = 29.63$ (Radial acceleration: Start from C and go in the direction of C to D).
8. $c_2 = $ Unknown (Tangential acceleration: In the direction of velocity of C w.r.t. D).
9. c is the intersection of c_1 and c_2.
10. $a,d\text{–}c$ is the acceleration of C $= 116.849$ (see Figure 12.93).
    (Figure 12.94 shows the combination of the space diagram, velocity and acceleration diagrams.)

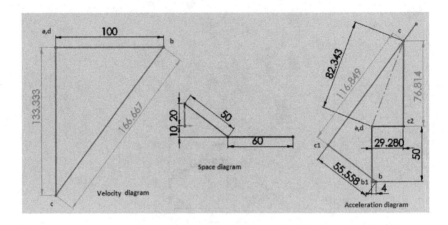

Fig. 12.94 Completed velocity and acceleration diagrams.

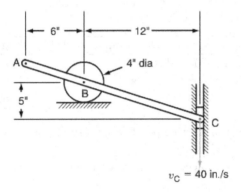

Fig. 12.95 Problem definition.

Problem 12.11

For the system shown in Figure 12.95 determine (a) the angular velocity of AC, (b) the velocity of point A, and (c) the angular velocity of the roller.

SolidWorks solution

The sketch of the mechanism is first produced as shown in Figure 12.96. The velocity diagram is the drawn using SolidWorks Sketching Tools. The surface supporting roller B and the slot in which C moves are considered as belonging to the body of the mechanism or machine (grounded).

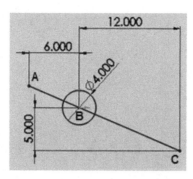

Fig. 12.96 Sketch of the mechanism.

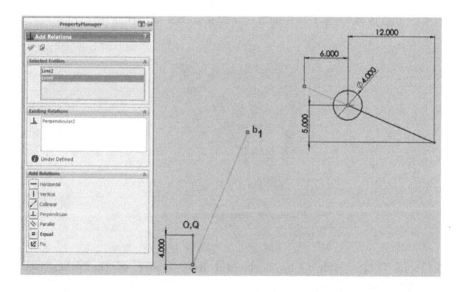

Fig. 12.97 Velocity of point C relative to the slot O.

The velocity diagram is the drawn using SolidWorks Sketching Tools. The horizontal and inclined hatched planes are considered as belonging to the body of the mechanism or machine (grounded).

1. Choose O, Q as the starting point (arbitrary in space) (see Figure 12.97).
2. Choose a scale of 1:10.
3. $Oc = 4\,inch/s$ ↓ (O–c is the velocity of C; given as being downward).
4. Sketch a line c–b_1 perpendicular to AB (see Figure 12.97).

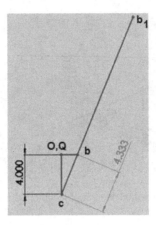

Fig. 12.98 Velocity of point B relative to surface supporting roller, Q.

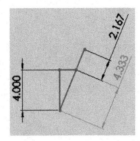

Fig. 12.99 Velocity of point A.

5. From O,Q sketch Q–b_2 horizontal from centre of the roller (see Figure 12.98).
6. b is the intersection of c–b_1 and Q–b_2 (see Figure 12.98).
7. c–b is the velocity of CB = 4.333 (taking the scale into consideration).
8. Using ratio, **a–b** is the linear velocity of AB = 2.167 (taking the scale into consideration) (see Figure 12.99).

The calculation for AB, BC, and AC are based on Figure 12.100.
From ratio, $\frac{Y}{18} = \frac{5}{12}$; $\therefore Y = 7.5$

$$AB = \sqrt{(7.5 - 5)^2 + 6^2} = 6.5$$

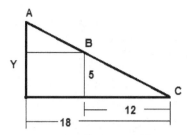

Fig. 12.100 Geometric calculation for dimensions.

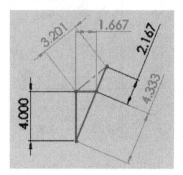

Fig. 12.101 Velocity diagram.

$$BC = \sqrt{5^2 + 12^2} = 13$$
$$AC = 6.5 + 13 = 19.5$$

The linear velocity of AC is obtained as follows using ratio:

$$\frac{ab}{bc} = \frac{AB}{BC}; \quad \therefore \ ab = bc \times \frac{AB}{BC} = 4.333 \times \frac{6.5}{13} = 2.167$$
$$\therefore \ ac = ab + bc = 04.333 + 2.167 = 6.5$$

Applying the scale used, ac = 65 m/s. Angular velocity of AC = linear velocity of AC/length of AC = $\frac{65}{19.5}$ = 3.33 rad/s. Velocity of A = velocity of A relative to O = 32.01 in/s (see Figure 12.101). Angular velocity of roller, $\omega_{\text{roller}} = \frac{ob}{\text{radius of roller}} = \frac{1.67}{2} = 0.835$ rad/s. Applying scale, angular velocity of roller = 8.35 rad/s.

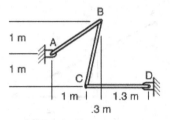

Fig. 12.102 Problem definition.

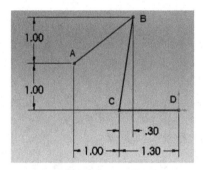

Fig. 12.103 Sketch of the mechanism.

Problem 12.12

Pin C of the linkage in Figure 12.102 has a velocity of 5 ms downward. Determine the velocity of B and the angular velocity of AB.

SolidWorks solution

The sketch of the mechanism is first produced as shown in Figure 12.103. The velocity diagram is drawn using SolidWorks Sketching Tools. Joints A and D are considered as belonging to the body of the mechanism or machine (grounded).

1. Choose A,D as the starting point (arbitrary in space) (see Figure 12.104).
2. Choose a scale of 1:10.
3. $D - c = 5\,m/s$ ↓ (D–c is the velocity of C; given as being downward).
4. Sketch a line c–b_1 perpendicular to BC (see Figure 12.104).
5. From A,D sketch A–b_2 perpendicular to AB (see Figure 12.105).
6. b is the intersection of c–b_1 and A–b_2 (see Figure 12.106).
7. a–b is the velocity of AB = 4.4 (taking the scale into consideration) (see Figure 12.106).

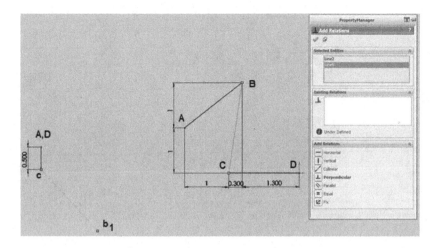

Fig. 12.104 Velocity of point B relative to point C.

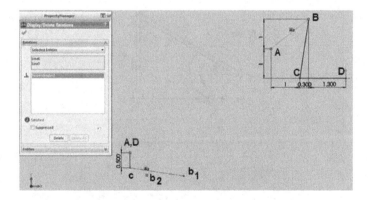

Fig. 12.105 Velocity of point B relative to point A.

Problem 12.13

Link BC in Figure 12.107 is pinned to a cylinder rolling at 12 in/s to the right. Calculate the velocity of pin C for the instant shown.

SolidWorks solution

The sketch of the mechanism is first produced as shown in Figure 12.108. The velocity diagram is the drawn using SolidWorks Sketching Tools. The

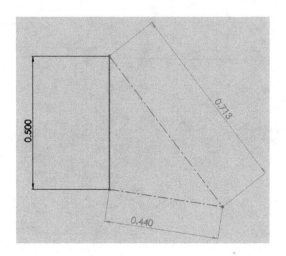

Fig. 12.106 Velocity diagram.

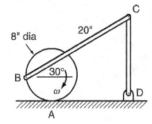

Fig. 12.107 Problem definition.

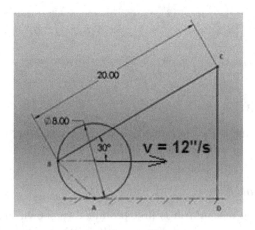

Fig. 12.108 Sketch of the mechanism.

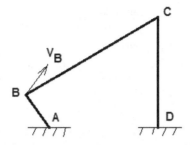

Fig. 12.109 Sketch of the equivalent mechanism.

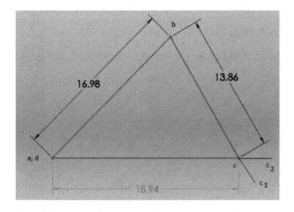

Fig. 12.110 Velocity diagram.

surface supporting roller B and joint D are considered as belonging to the body of the mechanism or machine (grounded).

The mechanism becomes a four-bar linkage with members being AB, BC, CD, and the frame AD (immovable). Figure 12.109 shows a clearer representation of the mechanism for analysis.

AB is the input while CD is the output.

$$AB = \sqrt{4^2 + 4^2} = 5.66''$$

At point A,

$$\frac{v}{r} = \frac{(v_B)_A}{AB}$$

$$\therefore (v_B)_A = AB \times \frac{v}{r} = 5.66 \times \frac{12}{4} = 16.98''/s$$

1. Choose a,d as the starting point (arbitrary in space) (see Figure 12.110).
2. Choose a scale of 1:1.

3. From a,d sketch a line perpendicular to AB, equal to 16.98.
4. From b sketch a line b–c_1 perpendicular to BC.
5. From a,d sketch d–c_2 perpendicular to DC.
6. b is the intersection of b–c_1 and d–b_2 (see Figure 12.110).
7. d–c is the velocity of C = 18.94 (i.e. 18.94 in/s to the right).

Problem 12.14

Member AC rotates at 2 rad/s counter-clockwise (Figure 12.111). Determine the velocity of point D, angular velocity of DC, and velocity of point G.

SolidWorks solution

The first step is to determine the length of CB using Figure 12.112.

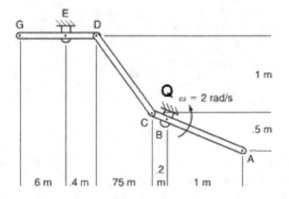

Fig. 12.111 Problem definition.

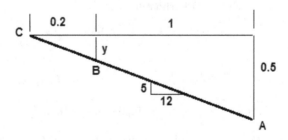

Fig. 12.112 Length of CB.

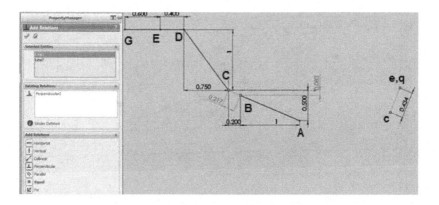

Fig. 12.113 Velocity of member AC.

Using ratio,

$$\frac{y}{0.5} = \frac{0.2}{(1+0.2)} = \frac{0.2}{1.2}$$

$$\therefore \; y = 0.5 \times \frac{0.2}{1.2} = 0.0833$$

$$\therefore \; CB = \sqrt{0.0833^2 + 0.2^2} = 0.217$$

The velocity of CA (CB), acting perpendicular to CA downward is:

$$v_C = r\omega = 0.217 \,(2\,\mathrm{rad/s}) = 0.433\,\mathrm{rad/s}$$

The sketch of the mechanism is first produced as shown in Figure 12.113. The velocity diagram is the drawn using SolidWorks Sketching Tools. Fixtures E and Q are considered as belonging to the body of the mechanism or machine (grounded).

1. Choose e,q as the starting point (arbitrary in space) (see Figure 12.113).
2. Choose a scale of 1:1.
3. From e,q sketch a line perpendicular to BC, equal to 0.433 (q–c is velocity of C).
4. From c sketch a line c–d$_1$ perpendicular to BC (see Figure 12.114).
5. From e,q sketch e–d$_2$ perpendicular to ED.
6. d is the intersection of c–d$_1$ and e–d$_2$ (see Figure 12.115).

Figure 12.116 is the velocity diagram showing the following:

1. Velocity of point D = 0.275 m/s.
2. Angular velocity of DC, $\omega_{DC} = \frac{cd}{CD} = \frac{0.209}{1.25} = 0.167\,\mathrm{rad/s}$.
3. Velocity of G, $\frac{de}{ge} = \frac{DE}{GE}$; $\Rightarrow ge = \frac{GE}{DE} \times de = \frac{0.6}{0.4} \times 0.275 = 0.4125\,\mathrm{m/s}\uparrow$.

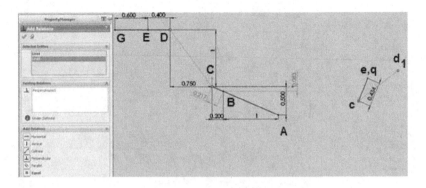

Fig. 12.114 Velocity of point C relative to point Q.

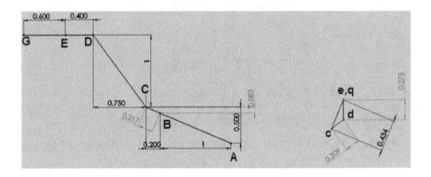

Fig. 12.115 Velocity of point D relative to point E.

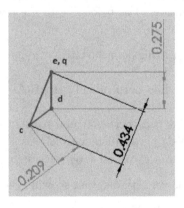

Fig. 12.116 Velocity diagram.

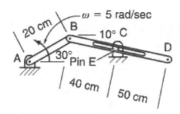

Fig. 12.117 Problem definition.

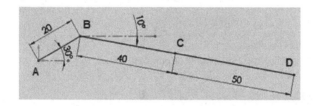

Fig. 12.118 Sketch of the mechanism.

Problem 12.15

Link AB of the mechanism shown in Figure 12.117 rotates at 5 rad/s counter-clockwise direction. Point C, a point on link BD, is assumed to be directly over pin e for the purposes of this problem. Determine the velocity of points B, C, and D, and the angular velocity of link BD.

SolidWorks solution

The sketch of the mechanism is first produced as shown in Figure 12.118. The velocity diagram is the drawn using SolidWorks Sketching Tools. Joints A and C are considered as belonging to the body of the mechanism or machine (grounded).

The velocity of B relative to A is: $(v_B)_A = r\omega = 20\,(5\ \text{rad/s}) = 100\,\text{rad/s}$.

1. Choose a,e as the starting point (arbitrary in space) (see Figure 12.119).
2. Choose a scale of 1:1.
3. Sketch a line a–b perpendicular to AB (see Figure 12.119).
4. Sketch a line b–c_1 perpendicular to BC (see Figure 12.120).
5. From **a,e** sketch e–c_2 parallel to BD (See Figure 12.121).
6. c is the intersection of b–c_1 and e–c_2 (see Figure 12.122).

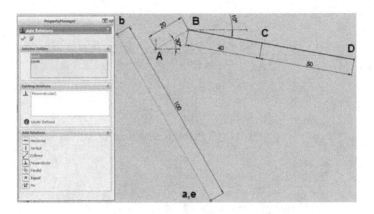

Fig. 12.119 Velocity of point B relative to point A.

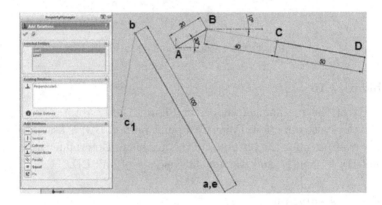

Fig. 12.120 Velocity of point C relative to point B.

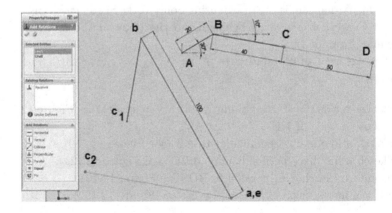

Fig. 12.121 Velocity of point C relative to point E (constrained along the slot).

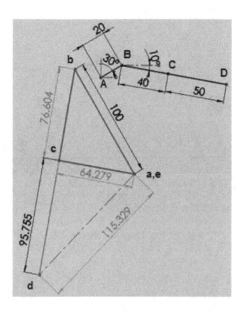

Fig. 12.122 Velocity diagram.

$$\frac{cd}{bc} = \frac{CD}{BC}; \quad \Rightarrow cd = bc \times \frac{CD}{BC} = 76.604 \times \frac{50}{40} = 95.755 \, \text{cm/s}$$

From the velocity diagram of Figure 12.122, we can glean all the solutions:

$$(v_B)_A = 100 \, \text{cm/s}$$
$$(v_C)_E = 64.279 \, \text{cm/s}$$
$$(v_D)_E = 115.33 \, \text{cm/s}$$
$$(\omega)_{BD} = \frac{(95.755 + 76.604)}{(40 + 50)} = 1.915 \, \text{rad/s}$$

Problem 12.16

Determine the angular acceleration of the link CD for the case shown in Figure 12.123 in which the input AB rotates with an angular velocity of 480 rad/s.

SolidWorks solution

The sketch of the mechanism is first produced as shown in Figure 12.124. The velocity diagram is drawn using SolidWorks Sketching Tools. Joints A

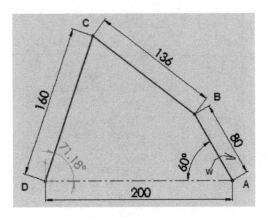

Fig. 12.123 Four-bar mechanism.

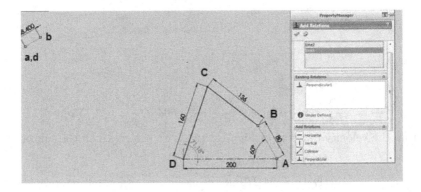

Fig. 12.124 Velocity of point B with respect to point A.

and D are considered as belonging to the body of the mechanism or machine (grounded).

The velocity of B relative to A is: $(v_B)_A = \omega \times AB = 480 \times 0.08 = 38.4$

1. Choose a,d as the starting point (arbitrary in space) (see Figure 12.124).
2. Choose a scale of 1:1.
3. Sketch a line a–b perpendicular to AB (see Figure 12.124).
4. Sketch a line b–c_1 perpendicular to BC (see Figure 12.125).
5. From **a,d** sketch d–c_2 perpendicular to BC (See Figure 12.126).
6. c is the intersection of b–c_1 and d–c_2 (see Figure 12.127).

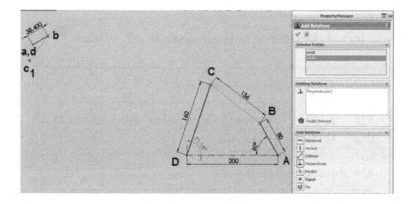

Fig. 12.125 Velocity of point C with respect to point B.

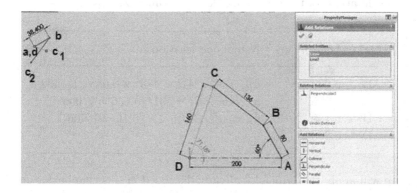

Fig. 12.126 Velocity of point C with respect to point D.

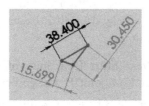

Fig. 12.127 Velocity diagram.

Velocity

The velocities of the different links are summarized as follows:

Link	Length (m)	Velocity (m/s)
AB	0.080	$(v_B)_A = \omega \times AB = 480 \times 0.08 = 38.4$
BC	0.136	30.45
CD	0.160	15.7
AD	0.200	0

Acceleration

The accelerations of the different links are summarized as follows:

Link	Length (m)	Radial acceleration (m/s^2)
AB	0.080	$a_R = \omega^2 \times AB = 480^2 \times 0.08 = 18,432$
BC	0.136	$a_R = bc^2/BC = 31^2/0.136 = 7,066$
CD	0.160	$a_R = cd^2/CD = 15^2/0.16 = 1,406$
AD	0.200	0

We will apply the principles already discussed in this chapter for accelerations diagrams.

Rule for acceleration diagram:

1. Start with radial acceleration (obtained from velocity diagram).
2. Follow with tangential acceleration (if it exists).
3. Continue from joint to joint.

Rule for radial acceleration: Start from joint of interest and go beyond where it is attached (start from joint B and go beyond A).

Rule for tangential acceleration: It is in the direction of the velocity.

In this problem, there is no initial tangential acceleration (it does not exist). Let us now start with radial acceleration at joint B towards joint A.

1. From a,d draw a line away in the direction from B with a value of 184, 320 (see Figure 12.128) (radial acceleration of B with respect to A; there is no tangential acceleration).

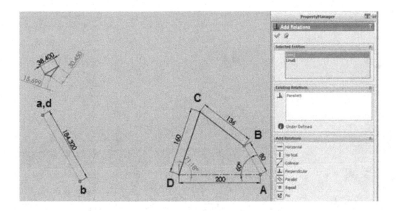

Fig. 12.128 Radial acceleration of point B with respect to point A.

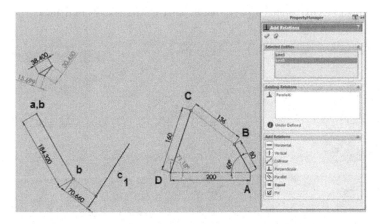

Fig. 12.129 Radial acceleration of point C with respect to point B and unknown tangential acceleration.

2. From **b**, draw a line away in the direction from C with a value of 70, 660 (see Figure 12.129) (radial acceleration of C with respect to B).

3. Draw an infinite length of line at right-angle to the preceding line (unknown tangential acceleration); call this line c_1 (see Figure 12.129).

4. From **c**, draw a line away in the direction from **a, d** with a value of 1,406 (see Figure 12.130) (radial acceleration of C with respect to D).

5. Draw and infinite length of line at right-angle to the preceding line (unknown tangential acceleration); call this line c_2 (see Figure 12.131).

6. The intersection of c_1 and c_2 gives the point c, for the acceleration of C.

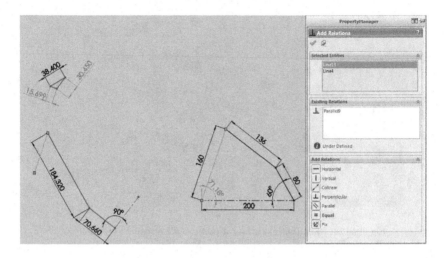

Fig. 12.130 Radial acceleration of point C with respect to point D.

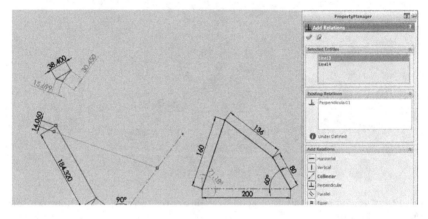

Fig. 12.131 Radial acceleration of point C with respect to point D and unknown tangential acceleration.

Problem 12.17

This is same arrangement with the tutorial in Problem 12.16 except that the link AB is decelerating at 80000 rad/s (i.e. in an anticlockwise direction). Determine the acceleration of the link CD.

SolidWorks solution

Follow the preceding step (see Figure 12.132); the solution is given in Figure 12.133.

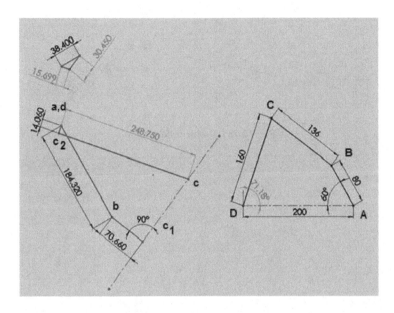

Fig. 12.132 Acceleration diagram (bottom left); velocity diagram (top left).

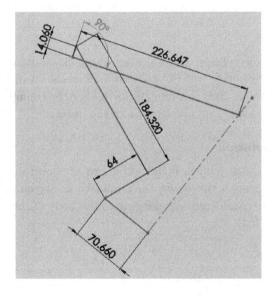

Fig. 12.133 Acceleration diagram.

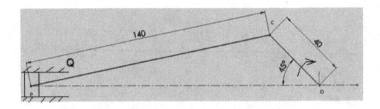

Fig. 12.134 Problem definition.

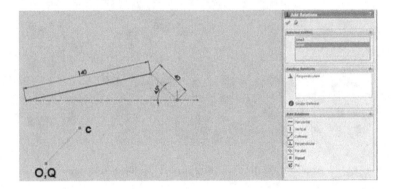

Fig. 12.135 Relative velocity of joint C with reference to joint O.

Problem 12.18

A horizontal single cylinder reciprocating engine has a crank OC of radius 40 mm and a connecting rod PC 140 mm long as shown in Figure 12.134. The crank rotates at 3000 rev/min clockwise. For the configuration shown, determine the velocity and acceleration of the piston.

SolidWorks solution

The sketch of the mechanism is first produced as shown in Figure 12.135. The velocity diagram is the drawn using SolidWorks Sketching Tools. Joints O and slot Q are considered as belonging to the body of the mechanism or machine (grounded).

Velocity Diagram

The velocity of C relative to A is:

$$\omega = \frac{2\pi(3000)}{60} = 314.16 \, \text{rad/s}$$

$$(v_B)_A = \omega \times AB = 314.16 \times 0.04 = 12.57 \, m/s$$

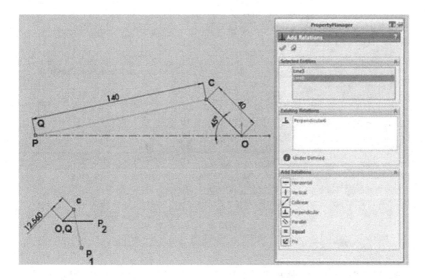

Fig. 12.136 Relative velocity of joint P with reference to joint C.

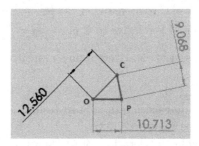

Fig. 12.137 Velocity of the piston = 10.713 m/s.

1. Choose O, Q as the starting point (arbitrary in space) (see Figure 12.135).
2. Choose a scale of 1:100.
3. Sketch a line O–c perpendicular to OC (see Figure 12.135).
4. Sketch a line c–p_1 perpendicular to CP (see Figure 12.136).
5. From **O, Q** sketch O–p_2 parallel to OP (see Figure 12.136).
6. p is the intersection of c-p_1 and O–p_2 (see Figure 12.136).

Figure 12.137 shows that the velocity of the piston = 10.713 m/s.

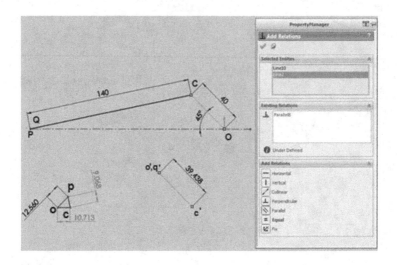

Fig. 12.138 Radial/centripetal acceleration of point C with respect to point O.

Acceleration diagram

The radial acceleration of C relative to O is: $= \frac{oc^2}{OC} = \frac{12.57^2}{0.04} = 39.438 \, \text{rad/s}^2$.

The radial acceleration of P relative to C is: $= \frac{cp^2}{CP} = \frac{9.068^2}{0.14} = 5.873 \, \text{rad/s}^2$.

1. Choose o', q' as the starting point (arbitrary in space) (see Figure 12.138).
2. Choose a scale of 1:100.
3. $o'-c' = 39.438$ (Radial acceleration: Start from C and go in the direction of joint C to O) (see Figure 12.138).
4. $c'-p' = 5.873$ (Radial acceleration: Start from P and go in the direction of P to C).
5. p'_1- = Unknown (Tangential acceleration: In the direction of velocity of P w.r.t. C) (see Figure 12.139).
6. $o-p'_2$ is the constraint for joint P to move only horizontally.
7. p is the intersection of tangential accelerations.
8. $o'-p$ is the acceleration of P = 2370.2 (see Figure 12.140).

Acceleration of the piston $= 23.702 \times 100 = 2370.2 \, \text{m/s}^2$.

Problem 12.19

Roller A of Figure 12.141 moves downward to the left at 45 in/s. Determine the linear velocity of point E.

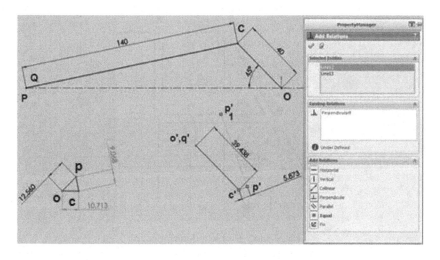

Fig. 12.139 Radial/centripetal acceleration of point P with respect to point C.

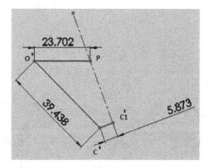

Fig. 12.140 Acceleration of the piston.

SolidWorks solution for velocity diagram

We will construct two lines: One from B to E and another from D to E (see Figure 12.142).

1. Choose a scale of 1:10.
2. Choose reference point (A, C, D) as the starting point (arbitrary in space).
3. From reference point, sketch velocity of A, $a = 45$ (a is velocity of A; down along $A–A_1$).

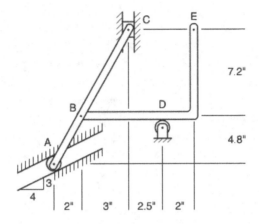

Fig. 12.141 Problem description.

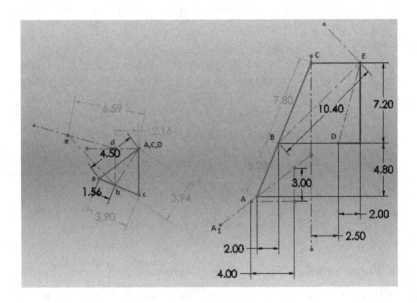

Fig. 12.142 Velocity diagram.

4. From the velocity, **a**, sketch **a–c** perpendicular to AC (velocity of C relative to A).
5. From the reference point, sketch a vertical line (velocity of C relative to frame is vertical).

6. Velocity of C, **c**, is the intersection of **ac** and vertical.
7. Use the Smart Dimension tool to dimension ac as 39 (The message 'Make Dimension driven?' appears. Accept 'Yes').
8. Velocity of D is perpendicular to BD from B; therefore sketch b–d vertical at d.
9. D is constrained to move horizontal by a roller so sketch a horizontal line from ACD.
10. Velocity of D is the intersection of b–d and D–d (see Figure 12.142).
11. Sketch a line d–e perpendicular to DE.
12. Sketch a line b–e perpendicular to BE.
13. Velocity of E is the intersection of d–e and b–e (see Figure 12.142).

Velocity of E is 65.9 in/s. This value matches exactly the value obtained analytically.

Problem 12.20

For the mechanism shown in Figure 12.143, determine:

1. The velocity of pin C.
2. The angular velocity of BCD.
3. The velocity of point D.

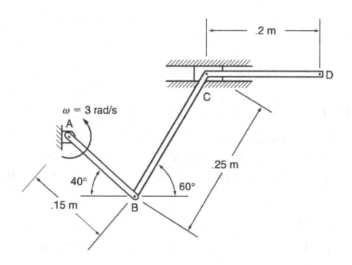

Fig. 12.143 Problem description.

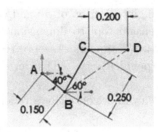

Fig. 12.144 Sketch of the mechanism.

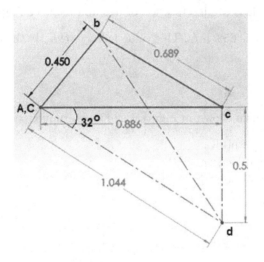

Fig. 12.145 Velocity diagram.

SolidWorks solution

The sketch of the mechanism is first produced as shown in Figure 12.144. The velocity diagram is the drawn using SolidWorks Sketching Tools. Joint A and slot C are considered as belonging to the body of the mechanism or machine (grounded).

Velocity diagram

The velocity of B relative to A is: $r\omega = 0.15(3 \text{ rad/s}) = 0.45 \text{ m/s}$.

1. Choose A,C as the starting point (arbitrary in space) (see Figure 12.145).
2. Choose a scale of 1:1.

3. Sketch a line A–b perpendicular to AB (see Figure 12.145).
4. Sketch a line b–c_1 perpendicular to BC (see Figure 12.145).
5. From **A,C** sketch C–c_2 parallel to slot (see Figure 12.145).
6. c is the intersection of b–c_1 and C–c_2 (see Figure 12.145).

1. The velocity of pin C = A,C-c = 0.886 m/s →.
2. The angular velocity of BCD = b-c/BC = 0.689/0.25 = 2.76 rad/s clockwise.
3. The velocity of point D is obtained by obtained as the intersection of a line from b perpendicular to BC and of a line from c perpendicular to CD. The length of the line from A,C to d gives the velocity of D = 1.044 m/s at an angle of 32° to the horizontal.

Problem 12.21

For the system shown in Figure 12.146, the velocity of point D is 51 in/s. Using the method of instantaneous center, determine:

1. The velocity of point A.
2. The angular velocity of BC.
3. The angular velocity of AD.

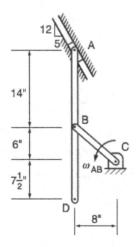

Fig. 12.146 Problem description.

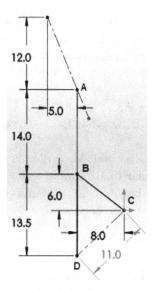

Fig. 12.147 Sketch of the mechanism.

SolidWorks solution

The sketch of the mechanism is first produced as shown in Figure 12.147. The velocity diagram is the drawn using SolidWorks Sketching Tools. Joint C and slot A are considered as belonging to the body of the mechanism or machine (grounded).

Figure 12.148 shows the instantaneous center radii for points A, B, and D.

$$\omega_{AD} = \frac{V_D}{r_D} = \frac{51}{25.5} = 2\,\text{rad/s}$$
$$V_A = \omega_{AD} \times r_A = 2(13) = 26\,\text{m/s}$$
$$V_B = \omega_{AD} \times r_B = 2(15) = 30\,\text{m/s}$$
$$\omega_{BC} = \frac{V_B}{BC} = \frac{30}{10} = 3\,\text{rad/s}$$

Problem 12.22

Determine the angular velocity of CBD and the linear velocity of D for the mechanism shown in Figure 12.149.

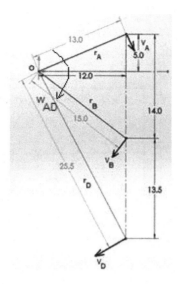

Fig. 12.148 Instantaneous center radii for points A, B, and D.

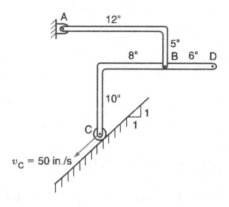

Fig. 12.149 Problem description.

SolidWorks solution

The sketch of the mechanism is first produced as shown in Figure 12.150. The velocity diagram is the drawn using SolidWorks Sketching Tools. Joints **A** and slope Q, (on which the roller C rolls) are considered as belonging to the body of the mechanism or machine (grounded).

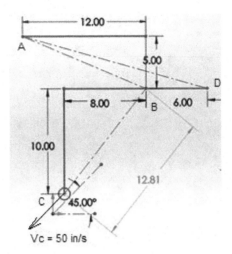

Fig. 12.150 Sketch of the mechanism.

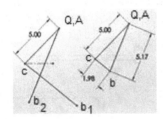

Fig. 12.151 Velocity diagram.

Velocity diagram

1. Choose A,Q as the starting point (arbitrary in space) (see Figure 12.151).
2. Choose a scale of 1:10.
3. Sketch a line A–c parallel to the incline plane (see Figure 12.151).
4. Sketch a line c–b_1 perpendicular to BC (see Figure 12.151).
5. From **A,Q** sketch A–b_2 perpendicular to AB (see Figure 12.151).
6. c is the intersection of c–b_1 and A–b_2 (see Figure 12.151).

Using the scale, we find from Figure 12.151 that

$$V_{B/C} = 19.8\,m/s; \quad V_B = 51.7\,m/s$$

$$\omega_{BC} = \frac{V_{B/C}}{|BC|} = \frac{19.8}{\sqrt{10^2 + 8^2}} = 1.54\,rad/s$$

$$V_{D/B} = \omega_{BC}|BD| = 1.54\,(6) = 9.3\,m/s$$

Velocity of D is obtained by the following steps:

1. From point b draw a vertical line = 0.93 in length (at right angle to BD) (note: We use the scale of 1:10 to sketch the line of length, 9.3/10 = 0.93).
2. From Q–A draw a line perpendicular to AD.

The intersection of these two lines define the point D (see Figure 12.152).

Problem 12.23

Determine the angular velocity of CBD and the linear velocity of D for the mechanism shown in Figure 12.153.

SolidWorks solution

The sketch of the mechanism is first produced as shown in Figure 12.154. The velocity diagram is the drawn using SolidWorks Sketching Tools. Slots

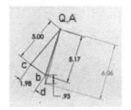

Fig. 12.152 Velocity diagram.

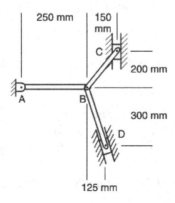

Fig. 12.153 Problem description.

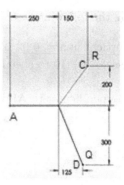

Fig. 12.154 Sketch of the mechanism.

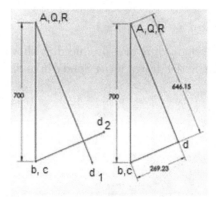

Fig. 12.155 Velocity diagram.

R, Q and joint A, are considered as belonging to the body of the mechanism or machine (grounded). Note: Block C slides on R while block D slides on Q.

Velocity diagram

1. Choose R, Q, and A as the starting point (arbitrary in space) (see Figure 12.155).
2. Choose a scale of 1:1000.
3. Since the velocity of block C slides vertically on Q, then the velocity of AB which is parallel to slot Q is the same as V_C, i.e. as $V_B = V_C$.
4. For velocity of D, sketch a line A-Q-R vertically downward to correspond to $V_C = 0.7\,\mathrm{m/s}$.
5. Sketch a line A–Q–R–d_1 parallel to BD parallel to slot (see Figure 12.155).

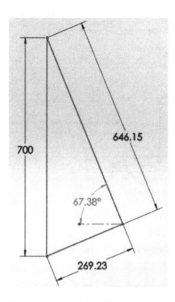

Fig. 12.156 Direction of the velocity.

6. Sketch a line $c,b-b_2$ perpendicular to BD (see Figure 12.155).
7. **d** is the intersection of $A-Q-R-d_1$ and $b,c-d_2$ (see Figure 12.155).

Using the scale, 1:1000 the velocity of D is $V_B = 646.15/1000 = 0.646$ m/s. The direction of the velocity is easy to obtain from the diagram as shown in Figure 12.156; it is $67.38°$ to the horizontal.

Problem 12.24

For the mechanism shown in Figure 12.157 (all dimensions in mm), C has a velocity of 1.25 m/s up the slope. Determine:

1. The angular velocity of CBD.
2. The linear velocity of D.

SolidWorks solution

The sketch of the mechanism is first produced as shown in Figure 12.158. The velocity diagram is the drawn using SolidWorks Sketching Tools. Joints **A** and slope Q (on which the roller C rolls) are considered as belonging to the body of the mechanism or machine (grounded).

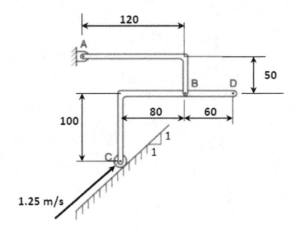

Fig. 12.157 Problem description.

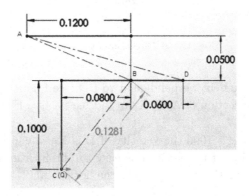

Fig. 12.158 Sketch of the mechanism.

Velocity diagram

1. Choose A, Q as the starting point (arbitrary in space) (see Figure 12.159).
2. Choose a scale of 1:10.
3. Sketch a line A–c parallel to the incline plane (see Figure 12.159).
4. Sketch a line c–b_1 perpendicular to BC (see Figure 12.159).
5. From **A, Q** sketch A–b_2 perpendicular to AB (see Figure 12.159).
6. c is the intersection of c–b_1 and A–b_2 (see Figure 12.159).

Using the scale, we find from Figure 12.160 that

$$V_{B/C} = 0.495 \, \text{m/s}; \quad V_B = 1.293 \, \text{m/s}$$

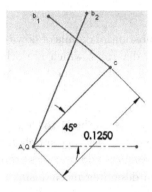

Fig. 12.159 Velocity diagram.

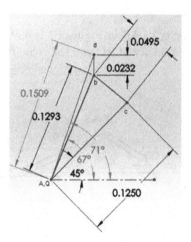

Fig. 12.160 Velocity diagram.

$$\omega_{BC} = \frac{V_{B/C}}{|BC|} = \frac{0.495}{\sqrt{0.8^2 + 0.1^2}} = 3.865 \, \text{rad/s}$$

$$V_{D/B} = \omega_{BC}|BD| = 3.865 \, (0.06) = 0.232 \, \text{m/s}$$

Velocity of D is obtained by the following steps:

1. From point b, draw a vertical line = 0.232 in length (at right angle to BD) (note: We use the scale of 1:10 to sketch the line of length: 0.232/10 = 0.0232).

2. From Q–A draw a line perpendicular to AD.

The intersection of these two lines define the point D (see Figure 12.160).

Summary

This chapter has given substantial details of how to use SolidWorks to draw velocity and acceleration diagrams which are otherwise more challenging and time consuming using analytical methods. Acceleration diagrams are extremely important because the force that a moving body has is dependant on its mass (or moment of inertia) and the total accelerations (tangential and normal). Using analytical methods to determine acceleration is very challenging but SolidWorks offers a very simple method using velocity and acceleration diagrams. Designers will find these tools very useful during the initial design phase when designing mechanisms because they can be first checked for dynamic performances before going into detail design.

Exercises

P1. The angular velocity of AB in Figure P1 is 400 rpm clockwise. Determine the angular velocity of BC and the velocity of C when (a) $\theta = 0°$ and $\theta = 90°$.

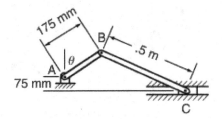

Fig. P1

P2. The velocity of slider C is 26 in/s downward (Figure P2). Determine the linear velocity of B and the angular velocity of AB.

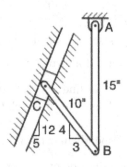

Fig. P2

P3. At the instant shown in Figure P3, AB has an angular acceleration of 8 rad/s² clockwise. Determine the acceleration of C and the angular acceleration of BC if AB has an angular velocity of 3 rad/s clockwise.

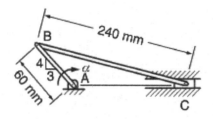

Fig. P3

P4. The input arm O_1A rotates 10 rad/s clockwise in Figure P4. Determine the velocity of the midpoint of AC.

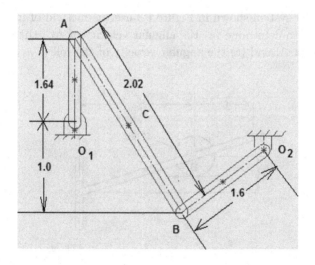

Fig. P4

P5. The input arm OA rotates 5 rad/s clockwise in Figure P5. Determine the velocity of the midpoint of AC.

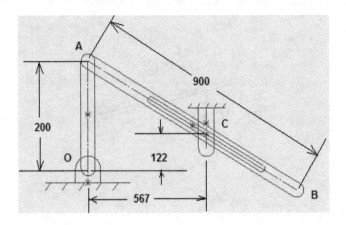

Fig. P5

P6. For the system shown in Figure P6 use the method of instantaneous center to determine (a) the angular velocity of AC, (b) the velocity of point A, and (c) the angular velocity of the roller.

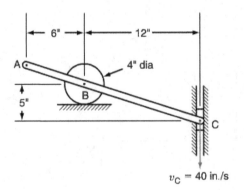

Fig. P6

P7. Repeat Problem 12.22 with the velocity of roller C upward along the slope.

P8. Repeat Problem 12.23 with the link AB inclined at 15° to the horizontal shown in Figure P7.

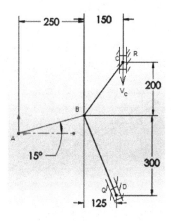

Fig. P7

P9. The angular velocity of link DE is 5 rad/s anti-clockwise. Determine the angular velocity of DB at the instant shown in Figure P8.

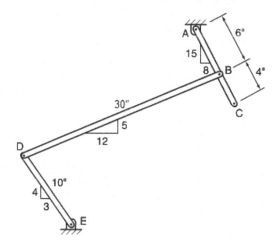

Fig. P8

P10. Pin B of the linkage has a velocity of 15 m/s to the right at the instant
   shown in Figure P9. Determine:

   (a) The angular velocity of pin C.
   (b) The angular velocity of AC.
   (c) The angular velocity of BD.

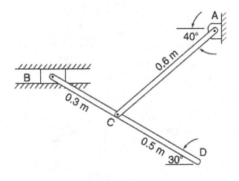

Fig. P9

Reference

Walker, K. M., *Applied Mechanics for Engineering Technologists*, 8th Edition,
   Prentice Hall, Upper Saddle River, NJ, 2007.

Chapter 13

Kinematics: Plane Motion of Mechanisms II

Objectives: When you complete this chapter you will:

- Have understood the concept of using SolidWorks to draw velocity diagrams.
- Have learnt how to solve problems related to plane motion of mechanisms using velocity diagrams.

Introduction: The Velocity Diagram Method

This chapter involves the construction of *velocity diagrams* which need to be done accurately and to a suitable scale. The normal practice is that students and practitioners use a drawing board, ruler, compass, protractor, and triangles and should possess the necessary drawing skills. This chapter introduces SolidWorks as a useful CAD tool to replace the out-dated approach of drawing board and the associated tools above.

The most intuitive and easiest way of dealing with most mechanisms when plane motion is involved is to use *velocity diagram*. This method can deal with both velocity and acceleration much more easily than the *relative motion* method.

Drawing velocity diagrams relies on the fact that the tangential velocity of a link is perpendicular to the link. If a link rotates with an angular velocity, then the tangential velocity is the product of the angular velocity and the length of the link. Mathematically stated this means

$$v = \omega \times r$$

Where the links are connected to the chassis or body of the machine, then those points are labeled using upper case. The values of the velocity are designated as lower cases.

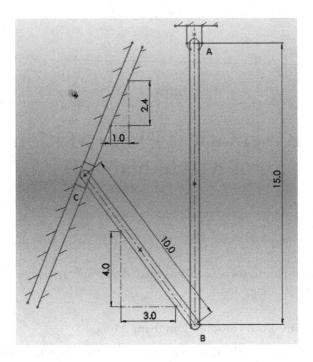

Fig. 13.1 Problem definition.

Example 13.1

The velocity of slider C is 26 in/s downward (Figure 13.1). Determine (a) the linear velocity of B and (b) the angular velocity of AB.

Solution of Example 13.1 using velocity diagram

We will redraw Figure 13.1 in a schematic form shown in Figure 13.2 to speed up the process although this step is not entirely necessary since we can still work with Figure 13.1. Then, we go through the following steps:

1. Start a New SolidWorks Part document.
2. Points Q and A are where the *slot* and *link AB* are fixed to the body of the machine; these are reference points and should be in capital letters. As already discussed, we note that velocities are presented using lower case.
3. The problem states that slider C is 26 in/s downward. Therefore, from A, Q sketch a line up to point *c* at a distance of 26 *parallel* to the slot (see

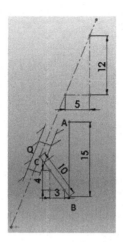

Fig. 13.2 Schematic of the given problem in Figure 13.1.

Figure 13.3). This means that you must work with the given diagram of the mechanism (Figure 13.1).

4. From point *c* we want to locate the velocity of B. Sketch a line from point *c* perpendicular to CB (because velocity is perpendicular to a link). Do not worry about how long the line is; this is one of the reasons this method is very easy to use.

5. We do not know the velocity of the point B on the mechanism yet, because point B is not only connected to link CB, it is also connected to link AB. So, from point A, Q sketch a line perpendicular to link AB. Where this line intersects the one drawn in Step 3 above, gives the velocity *b* (see Figure 13.3 for the solution).

6. Use the **Smart Dimension** tool to measure the distance A, Q to b and c to b. The first gives the velocity of B, while the second gives the velocity of B relative to C. Therefore, the answer to (a) is 22 in/s ($\rightarrow$) while the angular velocity of AB is $\omega = 22/15 = 1.47$ rad/s (counter-clockwise). How do we know that the angular velocity is counter-clockwise? We know because if we were standing at A and watch C move downwards it will appear to be moving in a counter-clockwise manner.

Example 13.2

The velocity of slider C is 30 in/s upward (Figure 13.1). Determine (a) the linear velocity of B and (b) the angular velocity of AB.

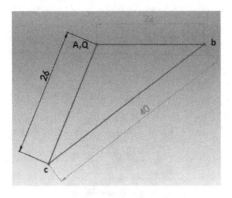

Fig. 13.3 Velocity diagram of Example 13.1.

Solution of Example 13.2 using velocity diagram

Let us go through the following steps:

1. Start a New SolidWorks Part document.
2. Points Q and A, are where the *slot* and *link AB* are fixed to the body of the machine; these are reference points and should be in capital letters. As already discussed, we note that velocities are presented using lower case.
3. The problem states that slider C is 30 in/s upward. Therefore, from A, Q sketch a line up to point *c* at a distance of 30 *parallel* to the slot (see Figure 13.4). This means that you must work with the given diagram of the mechanism (Figure 13.1).

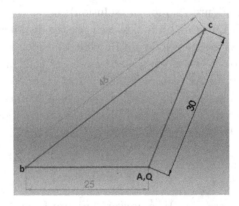

Fig. 13.4 Velocity diagram of Example 13.2.

4. From point *c* we want to locate the velocity of B. Sketch a line from point *c* perpendicular to CB (because velocity is perpendicular to a link). As above, do not worry about how long the line should be.

5. We do not know the velocity of the point B on the mechanism yet because point B is not only connected to link CB it is also connected to link AB. So, from point A, Q sketch a line perpendicular to link AB. Where this line intersects the one drawn in Step 3 above, gives the velocity *b* (see Figure 13.4 for the solution).

6. Use the **Smart Dimension** tool to measure the distance A, Q to b and c to b. The first gives the velocity of B, while the second gives the velocity of B relative to C. Therefore, the answer to (a) is 25 in/s ($\rightarrow$) and for (b) the angular velocity of AB is $\omega = 25/15 = 1.67$ rad/s (clockwise). How do we know that the angular velocity is clockwise? I know it is, because if we were standing at A and watch C move upwards, it will appear to be moving in a clockwise manner.

Example 13.3

Cylinder A (Figure 13.5) rolls down the slope with an angular velocity of 2 rad/s. Determine (a) the velocity of C and (b) the angular velocity of BC.

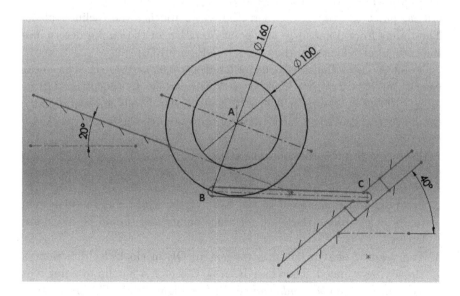

Fig. 13.5 Problem definition.

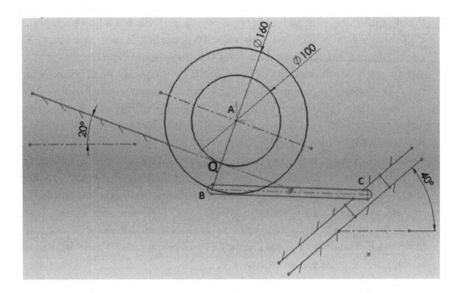

Fig. 13.6 The rolling wheel problem definition.

Solution of Example 13.3 using velocity diagram

This problem is referred to as the Rolling Wheel problem, in which the critical point to note is Q where the wheel touches the inclined plane. The radius from point Q to any point on either of the circles determines the tangential velocity on the circles (see Figure 13.6).

Point A has a linear velocity of $2\,\text{rad/s} \times 0.05 = 0.1\,\text{m/s}$. The linear velocity of B is $2\,\text{rad/s} \times 0.03 = 0.06\,\text{m/s}$ parallel to the plane; QB = $(160 - 100)/2 = 30\,\text{mm}$ or $0.03\,\text{m}$. Since the wheel rolls down the plane, the linear (tangential) velocity at B is tangential to QB in the clockwise direction.

Let us go through the following steps:

1. Start a New SolidWorks Part document.
2. Points Q and C are fixed to the body of the world coordinate system (body of machine); these are reference points and should be in capital letters.
3. The velocity of B is $0.06\,\text{m/s}$ tangent to QB in clockwise direction as already determined. Therefore, from Q, C sketch a line up to point *b* at a distance of 0.06 *perpendicular* to QB (see Figure 13.7). This means that you must work with the given diagram of the mechanism (Figure 13.6).

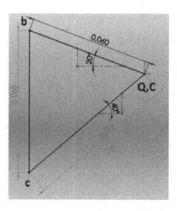

Fig. 13.7 Velocity diagram of Example 13.3.

4. From point **b** we want to locate the velocity of C on the slot. Sketch a line from point **b** perpendicular to BC (because velocity is perpendicular to a link). Do not worry about how long the line is.

5. We do not know the velocity of the point C on the mechanism yet because point C is not only connected to link BC it also slides along the slot. So, from point Q, C sketch a line parallel to the slot. Where this line intersects the one drawn in Step 4 above, gives the velocity **c** (see Figure 13.7 for the solution).

6. Use the **Smart Dimension** tool to measure the distance Q, C to c and c to b. The first gives the velocity of C, while the second gives the velocity of B relative to C. Therefore, the answer to (a) is 0.074 m/s (at 40° to the horizontal) and (b) the angular velocity of AB is $\omega = 0.068/0.2 = 0.34$ rad/s (clockwise).

Example 13.4

The cylinder shown in Figure 13.8 rolls to the right at 10 in/s. Determine the angular velocity of (a) BD and (b) CD.

Solution of Example 13.4 using velocity diagram

This problem illustrates how CAD can help reveal some errors in dimensions in machines when they are subjected to further analysis. The original problem from the reference source has the length of CD = 13 in. When this value was used in modeling the geometry shown in Figure 13.8,

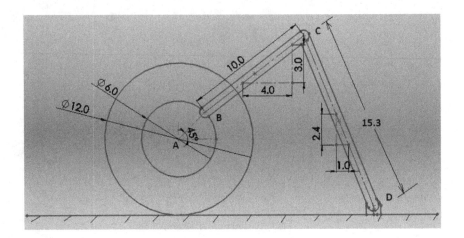

Fig. 13.8 Problem definition.

SolidWorks showed that there was inconsistency and gave the length of CD as 15.3 inches which is correct.

Let us analyze this given rolling wheel problem shown in Figure 13.8. We also note that the point B on the smaller circle makes an angle of 45° to the horizontal. Again, the critical point to note is Q where the wheel makes contact with the surface. We are interested in finding the radius of B from Q (the instantaneous centre). So we consider triangle BAQ and apply cosine rule, noting that AB = 6/2 = 3, AQ = 12/2 = 6 and the angle at A in triangle BAQ = 90 + 45 = 135°:

$$r_B^2 = 3^2 + 6^2 - 2(3)(6)\cos(135) = 70.45; \quad r_B = \sqrt{70.45} = 8.39$$

Our analysis results in an equivalent four-bar chain mechanism, QB, BC, CD, and QD. If the wheel rolls to the right, then it is rolling in a *clockwise* direction, and the velocity of A is to *right*.

$$\omega = \frac{v_A}{r_A} = \frac{v_B}{r_B}; \quad v_B = \frac{r_B}{r_A}v_A = \frac{8.39}{6}(10) = 13.98\,\text{in/s}$$

Let us go through the following steps:

1. Start a New SolidWorks Part document.
2. Points Q and D are fixed to the body of the world coordinate system (body of machine); these are reference points and should be in capital letters.

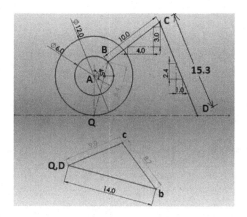

Fig. 13.9 Velocity diagram of Example 13.4.

3. The velocity of B is 13.98 in/s perpendicular to QB in clockwise direction as already determined. Therefore, from Q, D sketch a line up to point **b** at a distance of 13.98 *perpendicular* to QB (see Figure 13.9). This means that you must work with the given diagram of the mechanism (Figure 13.8).

4. From point **b** we want to locate the velocity of C. Sketch a line from point **b** perpendicular to BC (because velocity is perpendicular to a link). Do not worry about how long the line is.

5. We do not know the velocity of the point C on the mechanism yet, because point C is not only connected to link DC. So, from point Q, D sketch a line perpendicular to CD. Where this line intersects the one drawn in Step 4 above, gives the velocity **c** (see Figure 13.9 for the solution).

6. Use the **Smart Dimension** tool to measure the distance Q, D to c and c to b. The first gives the velocity of C, while the second gives the velocity of B relative to C. Therefore, the answer to (a) angular velocity of BC = 8.7/10 = 0.87 rad/s (counter-clockwise) and (b) the angular velocity of CD is $\omega = 9.0/15.3 = 0.58$ rad/s (clockwise).

Example 13.5

The angular velocity of CD is 3 rad/s for the walking link device shown in Figure 13.10. Determine (a) the velocity of point B, (b) the angular velocity of BD, and (c) the velocity of point E.

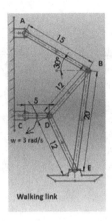

Fig. 13.10 Problem definition.

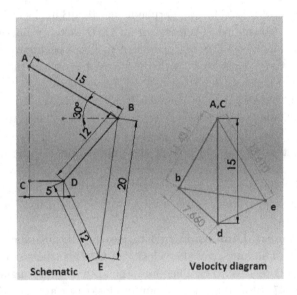

Fig. 13.11 Velocity diagram of Example 13.5.

Solution of Example 13.5 Using Velocity Diagram

We show the schematic of the given problem in Figure 13.11. Let us go through the following steps:

1. Start a New SolidWorks Part document.

2. Points A and C are fixed to the body of the world coordinate system (body of machine); these are reference points and should be in capital letters.

3. The velocity of D is 15 in/s perpendicular to CD in clockwise direction as already determined (acting downward). Therefore, from A, C sketch a line up to point *d* at a distance of **15** *perpendicular* to CD (see Figure 13.11). This means that you must work with the given diagram of the mechanism (Figure 13.10).

4. From point *d* we want to locate the velocity of B. Sketch a line from point *d* perpendicular to BD (because velocity is perpendicular to a link). Do not worry about how long the line is.

5. We do not know the velocity of the point B on the mechanism yet, because point B is not only connected to link AB. So, from point A,C sketch a line perpendicular to AB. Where this line intersects the one drawn in Step 4 above, gives the velocity *b* (see Figure 13.11 for the solution).

6. Use the **Smart Dimension** tool to measure the distance A, C to b, b to d, and A, C to e. The first gives the velocity of B, while the second gives the velocity of B relative to C. Therefore, the answer to (a) is velocity of B = 11.43 in/s (clockwise), (b) the angular velocity of BD is $\omega = 7.66/12 = 0.638$ rad/s (clockwise), and (c) velocity of E = 13.61 in/s (see Figure 13.11).

SolidWorks shows that the initial geometric model of Figure 13.10 of the problem definition is not consistent with the inclusion of 40° in the angle defining DBA. There is a geometric violation and therefore only the angle of 30° was necessary to be included since the software automatically computes the remaining angle.

Example 13.6

A piston containing a rod and crank mechanism is shown in Figure 13.12. The crank rotates at a constant velocity of 300 rad/s. Find (a) the acceleration of the piston and (b) the angular acceleration of the link BC.

SolidWorks Solution of Example 13.6 using velocity diagram

First calculate the tangential velocity of B relative to A: $(v_B)_A = \omega \times 0.05 = 15$ m/s.

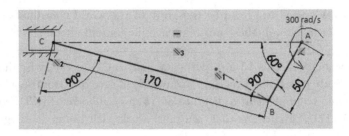

Fig. 13.12 Configuration of the problem to be solved.

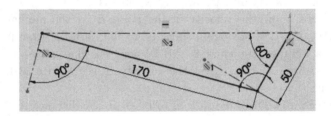

Fig. 13.13 Construction lines at B and C perpendicular to AB and BC respectively.

*Sketch the velocity diagram and determine the velocity
of C relative to B*

1. Create Sketch1 in SolidWorks, which consists of the linkages, AB, BC, and AC and dimension them as shown in Figures 13.12 and 13.13.
2. Create construction lines at B and C perpendicular to AB and BC respectively (see Figures 13.12 and 13.13; these are used for the directions of the velocities).
3. Draw *ab* parallel to the construction line at B, perpendicular to AB (see Figure 13.3).
4. Use relations tool to make these two lines to be parallel.
5. Use the Smart Dimension tool to dimension *ab* to be 15 (a scale of 10 is used for Figures 13.1 and 13.2 for consistency; remember to divide the solution by 10).
6. Draw *bc* parallel to the construction line at C, perpendicular to BC (see Figure 13.14).
7. Use relations tool to make these two lines to be parallel.
8. Draw *ac* parallel to AC since the piston C reciprocates along AC (see Figure 13.14).

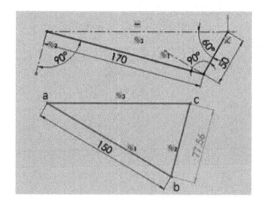

Fig. 13.14 Velocity diagram.

9. Use the Smart Dimension tool to dimension *bc* (you will receive a warning that this dimension is driven; accept this condition).

The driven dimension is automatically given as 77.56 by SolidWorks (see Figure 13.14).

Therefore, the velocity of BC is $(v_C)_B = 7.8\,\text{m/s}$ from SolidWorks.

Summary

This chapter has presented the Velocity Diagram method as an easy tool for solving some plane motion problems. SolidWorks presents very useful features for kinematics of mechanisms which are described as plane motion. When rolling wheels are involved, an initial pre-processing is necessary to reduce the problem to one that can be taken over by SolidWorks to solve the problem using the velocity diagram method. Not only are velocity diagrams easily handled, acceleration diagrams are similarly handled in such a user-friendly and easy way, thereby reducing analysis cycle time.

Reference

Walker, K. M., *Applied Mechanics for Engineering Technology*, 8th Edition, Prentice Hall, Upper Saddle River, NJ, 2007.

Chapter 14

Kinetics: Inertia Force and Torque

Objectives: When you complete this chapter you will have understanding on how to:

- Calculate the forces and acceleration of linear motion.
- Calculate the torque and acceleration of angular motion.
- Calculate the force, torque, linear acceleration, and angular acceleration of plane motion.
- Apply SolidWorks Motion to determine the forces and acceleration for linear, torque, and acceleration for angular, and force, torque, linear acceleration, and angular acceleration for plane motions.

Kinetics

Kinetics relates to the motion of material bodies and the forces associated therewith. Kinetics concerns not only velocity and acceleration but also the accompanying unbalanced forces that cause the motion.

Newton's first law of motion: Every object in a state of uniform motion tends to remain in that state of motion unless an external force is applied to it.

Newton's second law of motion: The relationship between an object's mass m, its acceleration a, and the applied force F is $F = ma$. Acceleration and force are vectors; in this law the direction of the force vector is the same as the direction of the acceleration vector.

Newton's third law of motion: For every action there is an equal and opposite reaction.

Newton's second law of motion

A body that has a resultant unbalanced force acting upon it behaves as follows:

1. The acceleration is proportional to the resultant force.
2. The acceleration is in the direction of the resultant force.
3. The acceleration is inversely proportional to the mass of the body.

Kinetics can be analyzed by three methods:

1. Inertia force or torque (dynamic equilibrium) — Chapter 14 of this book.
2. Work and energy — Chapter 15 of this book.
3. Impulse and momentum — Chapter 16 of this book.

This chapter will cover dynamic equilibrium as applied to linear, rotational, and plane motions using SolidWorks.

Linear Inertia Force

Newton's second law of motion can be represented in equation form. Let

$$a = \text{acceleration}$$
$$F = \text{resultant force}$$
$$m = \text{mass of body}$$
$$W = \text{force of gravity (or weight)}$$

1. $\mathbf{a} \; \alpha \; \mathbf{F}$ (or $a = (F)(\text{constant})$)
2. $\mathbf{a}$ and $\mathbf{F}$ are in the same direction
3. $\mathbf{a} \; \alpha \; \mathbf{1/m}$ (or $a = (\text{constant})/m)$)

The combination of these three statements is the equation:

$$\mathbf{F} = \mathbf{ma}$$

Where, in the SI system,

$$F = \text{force in Newton, N}$$
$$m = \text{mass in kilograms, kg}$$
$$a = \text{acceleration, m/s}^2$$

This equation is consistent with the original definition for $1\,\mathrm{N}$ being the force that causes a mass of $1\,\mathrm{kg}$ to accelerate at $1\,\mathrm{m/s}^2$

$$1 \text{ N} = (1 \text{ kg})(1 \text{ m/s}^2) \text{ or}$$
$$1 \text{ N} = 1 \text{ kg.m/s}^2$$

This force can also be the force of gravity on an object (customarily called weight). Taking the acceleration of gravity as $9.81\,\mathrm{m/s^2}$ and using W to represent weight, we get

$$\text{Weight} = (\text{mass})(\text{acceleration of gravity})$$

$$\mathbf{W} = \mathbf{mg}$$

If we have several forces causing a body to move with changing velocity (i.e. the body accelerates), we would have

$$\sum \mathbf{F} = \mathbf{ma}$$

Linear inertia force: Dynamic equilibrium

Recall Newton's third law: The opposing reaction or force is the inertia force, which is equal to ma. The inertia force will act opposite to the acceleration (see Figures 13.1–13.3).

- Ensure all forces on the FBD are included.
- Consider a force P that causes acceleration of the block to the right in Figure 14.1.
- The opposing force (or inertia force) is shown in the opposite direction to the acceleration vector in Figure 14.2.

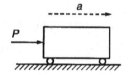

Fig. 14.1 Motion due to P.

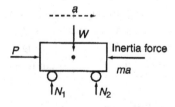

Fig. 14.2 Inertia force.

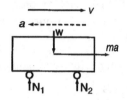

Fig. 14.3 Decerating block move right.

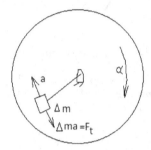

Fig. 14.4 Torque and angular acceleration.

To better illustrate that the inertia force always acts in a direction opposite to acceleration, consider the block moving to the right but decelerating in Figure 14.3.

Angular inertia torque: Dynamic equilibrium

This is the relation between torque and angular acceleration. Consider a mass m moving in a circle of radius r, acted on by a tangential force F_t as shown in Figure 14.4.

Applying Newton's second law to relate F_t to the tangential acceleration $a_t = r\alpha$, where α is the angular acceleration, $F_t = ma_t = mr\alpha$. The torque about the center of rotation due to F_t is $T = F_t \cdot r = mr^2\alpha$.

For a rotating rigid body made up of a collection of masses $m_1, m_2, \ldots$ the total torque about the axis of rotation is $T_{total} = \sum T_i = \sum(mr^2)$ $\alpha = I\alpha$, where the moment of inertia I of a rigid body gives a measure of the amount of resistance a body has to changing its state of rotational motion.

Note: The units of moment of inertia are kg.m^2 in SI and ft-lb-s^2 in the US customary system.

Plane motion inertia force and torque: Dynamic equilibrium

The combination of rectilinear or translation motion and centroidal rotation (angular) in one object gives plane motion such as a rolling wheel.

A rolling wheel:

- It travels down an incline.
- It also rotates about its axis.
- Two types of inertia are found.

Combining motion:

- Use both angular and linear motion.
- Linear (F = ma); angular (T = $I_c\alpha$).
- Linear forces and accelerations: x and y directions.
- Angular torques and accelerations: CW and CCW directions.

Dynamic equilibrium:
Inertial force and inertial torque are opposite to the direction of acceleration.

- Torque and force are used for calculating equilibrium.
- Use *ma* (in *x and y* directions) and $I\alpha$ for force and torque, respectively.

Rules for problem solving:

- Establish the direction of acceleration.
- Draw a FBD showing dynamic equilibrium, with the addition of either an inertia force and/or an inertia torque.
- Show the direction of acceleration beside the FBD.
- For linear motion, use $m \cdot a$, acting through the center of gravity, and opposite in direction to the acceleration a.
- For rotational motion, use $I\alpha$, opposite in direction to the angular acceleration α.

When using $I\alpha$ in a moment equation, remember that it is a moment, not a force.

In the remaining parts of this chapter, we will demonstrate how Solid-Works is used to solve problems in kinetics for the force/torque inertia domain. We will present the solution methodology for SolidWorks and then solve some problems and validate their solutions. Understanding the concepts involved will enhance using SolidWorks not only for design and analysis based on FEA but also for investigating dynamic machine elements.

SolidWorks Solution Procedure

1. Create a model of the **Part** being studied.
2. Create the **Part** model of *a body* on which the part being studied moves.
3. Create an **Assembly** of the part and the body.
4. Create **Mates** for the part and body (especially the **Limit Distance** mate to determine the motion constraints).
5. **Add-In** the **SolidWorks Motion** tool.
6. Select the **Motion Analysis** option from the left side of the **Motion-Manager** window.
7. From the **MotionManager**, click the **Gravity**, **Contact**, and **Force PropertyManagers** and define the necessary parameters (see Figure 14.5).
8. Click the **Calculate** button to calculate the **Motion Study**.
9. From the **MotionManager**, click the **Results and Plot** option and select the necessary options for display.

The tools shown above are the most commonly used tools for application in dynamics of machine members. The **Contact** tool is extremely useful for defining bodies that are in contact as well as the statics and kinetic coefficients of friction between the bodies. The **Force** tool is another useful tool which makes it easy for a user to define the surface on which force acts and its direction and value. The **Gravity** tool does not really need any further setting except to show its direction (it is normally downward). Figure 14.6 shows a typical Motion Study Manager obtained from a study.

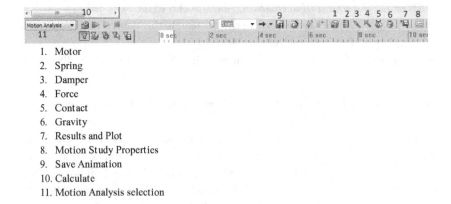

1. Motor
2. Spring
3. Damper
4. Force
5. Contact
6. Gravity
7. Results and Plot
8. Motion Study Properties
9. Save Animation
10. Calculate
11. Motion Analysis selection

Fig. 14.5 Motion Study interface.

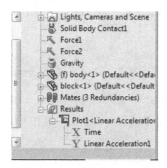

Fig. 14.6 Motion Study Manager.

Let us now consider some problems in order to understand how dynamics problems are handled using SolidWorks CAD software.

Two categories of problems are solved in this chapter: linear and plane motion. The first five problems have linear motion, while others have plane motion.

Kinetics: Linear Motion

Problem 14.1

Determine the acceleration of the 150 lb block shown in Figure 14.7 if the coefficient of kinetic friction is 0.4. $P_1 = 90$ lb; $P_2 = 60$ lb; $W = 150$ lb; $\theta_1 = 45°$; $\theta_2 = 30°$.

SolidWorks solution

Create a model of the **Part** being studied:

1. Open a **New SolidWorks Part** document.
2. Be in **Sketch** Mode, and select **Front Plane**.
3. Sketch a **Rectangle** 6 by 14.5 (unit is in inches) shown in Figure 14.8.
4. Extrude 6 in (see Figure 14.9).
5. Apply Material: AISI 304.

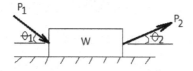

Fig. 14.7 Block on a horizontal plane acted upon by two forces.

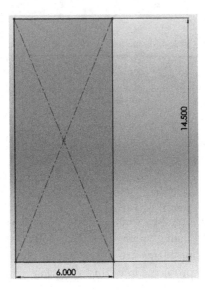

Fig. 14.8 Rectangular profile defining the block.

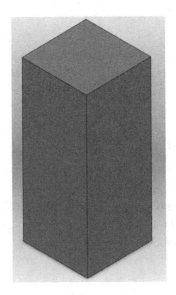

Fig. 14.9 Block.

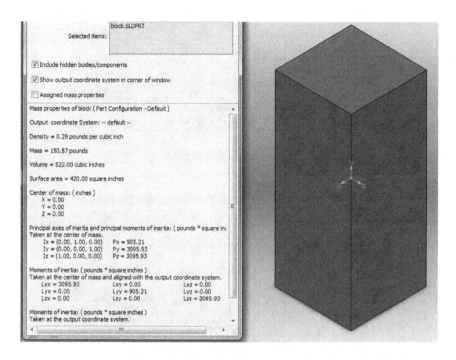

Fig. 14.10 Mass properties for the block.

6. Click **Evaluate > Mass Properties** (see Figure 14.10 for the mass properties).

Note that the dimensions have been so chosen to obtain a weight of 150 lb which is specified in the problem description. The choice of correct dimensions to obtain correct mass is an exercise for the student.

Create the **Part** model of a body on which the part being studied moves:

1. Open a **New SolidWorks Part** document.
2. Be in **Sketch** Mode, and select **Front Plane**.
3. Sketch the profile (unit is in inches) shown in Figure 14.11.
4. Extrude 6 in, Mid Plane (see Figure 14.12 for the Body).

The choice of material for the Body is not critical because using the **ContactManager** we can establish the coefficient of friction between the body and the block. However, the choice of material for the Block is important in order to obtain the specified mass. Once the Block and Body are

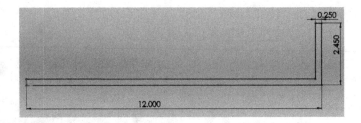

Fig. 14.11 Front profile of the body.

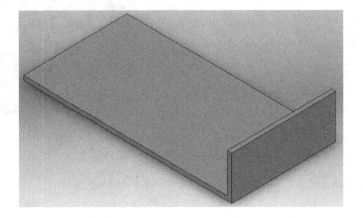

Fig. 14.12 Body.

modeled, the next step is to create an assembly of both and then proceed
with analysis.

Create an **Assembly** of the part and the body:

1. Open a **New SolidWorks Assembly** document.
2. Select the **Body** as the first component to **Insert** and click **OK** to
   accept it.
3. Select the **Block** as the second component to **Insert** and click **OK** (see
   Figure 14.13).

Create **Mates** for the part and body (especially the **Limit Distance**
mate to determine the motion constraints).

There are basically only three mates: *Two sides of the block* and *the
body* being **Coincident** and the **LimitDistance1** to limit the movement

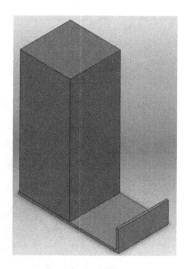

Fig. 14.13 Assembly of Body and Block.

of the block as shown in Figure 14.14. Figure 14.15 shows the **Assembly-Manager** with the three **Mates**.

Add-In the **SolidWorks Motion** tool:

1. Click **Add-In**.
2. Select the **SolidWorks Motion** tool (see Figure 14.16).

Select the **Motion Analysis** option:
Select the **Motion Analysis** option from the left side of the **MotionManager** window.

Activate the **Gravity, Contact,** and **Force PropertyManagers**:
Gravity

1. From the **MotionManager**, click the **Gravity** tool. The **Gravity PropertyManager** automatically appears (see Figure 14.18).
2. Click the **Y**-button (use the Gravity Parameter to choose the correct direction; an arrow shows up at the bottom right of the Graphics Window).
3. Click the **Check Mark** (click **OK** to finish Gravity definition).

Contact

4. From the **MotionManager**, click the **Contact** tool. The **Contact PropertyManager** automatically appears (see Figure 14.19).

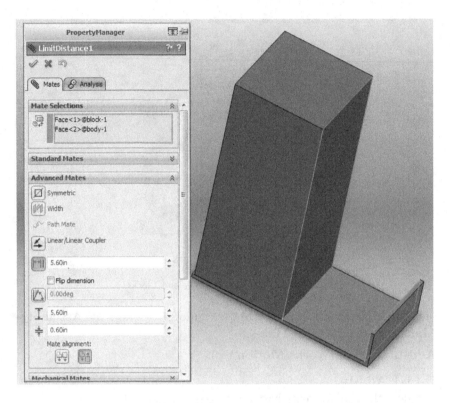

Fig. 14.14 LimitDistance1.

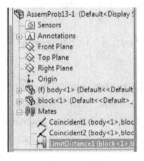

Fig. 14.15 AssemblyManager showing the three Mates.

Fig. 14.16 Add-In SolidWorks Motion.

Fig. 14.17 Selecting SolidWorks Motion.

5. **Uncheck** the **Material** option.
6. **Check** the **Friction** option.
7. Set **Dynamic Friction Velocity** $= 0$.
8. Set **Dynamic Friction Coefficient** $= 0.4$.
9. Set **Static Friction Velocity** $= 0$.
10. Set **Static Friction Coefficient** $= 0$.
11. In the **Elastic Properties** option, **check** the **Restitution** button.
12. Click **OK** to finish Contact definition.

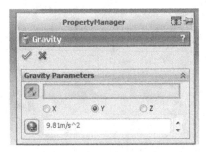

Fig. 14.18 Gravity PropertyManager.

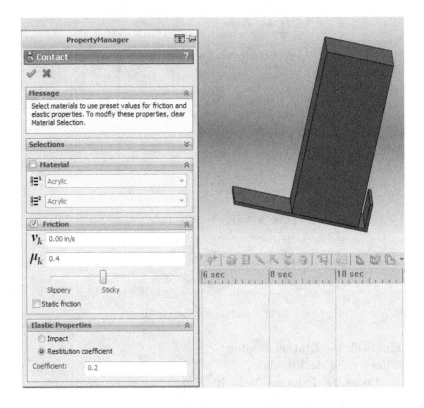

Fig. 14.19 Contact PropertyManager.

Force

Let us first review how SolidWorks deals with the force applied for dynamic analysis. After extensive simulation, the author found out that a preliminary analysis is required to obtain the force that is applied to the model being investigated. This requires the summation of forces in the direction motion of the body excluding the inertia force (ma). This summation of the forces is then applied to the body for simulation. Applying this preprocessing method to the problem at hand leads to the following.

The applied forces are resolved in both x- and y-direction. Figure 14.20 shows the FBD.

Applying equilibrium conditions:

$$\sum F_y = 0 : -63.6 - 150 + N + 30 = 0$$
$$N = 183.6$$
$$\therefore \ F = 0.4(183.6) = 73.44$$

$$\sum F_x = 0 : 63.6 + 52 - 73.44 = 42.16 \, \text{lb}$$

From the **MotionManager**, click the **Force** tool (see Figure 14.21).

13. For the **Direction**, select the *left face of the block* (**Face <1> @block-1**) as the **Action part and point of application of action** tool (if force is applied).
14. For the **Force Direction**, click the **Reverse Direction** if the need be to obtain correct direction.
15. For the **Force Function**, select the **Constant** for the pull-down menu.
16. For the **Force Value**, type the value of **42.16 lbf** and click **OK** to finish Force definition.

Calculate the **Motion Study**:
Click the **Calculate** button to calculate the **Motion Study** (see Figure 14.22).

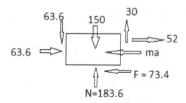

Fig. 14.20 FBD of force system.

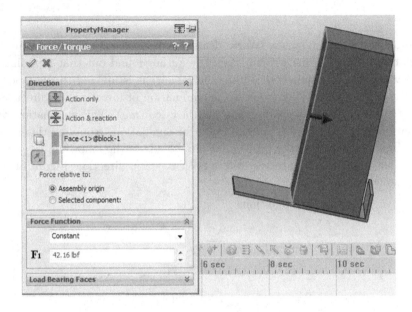

Fig. 14.21 Force/Torque PropertyManager.

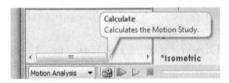

Fig. 14.22 Calculate the Motion Study.

Display the **Results and Plot**:
From the **MotionManager**, click the **Results and Plot** option (see Figure 14.23).

1. Right-click **Plot1** under **Results** folder (see Figure 14.24 for different results options).
2. Under **<Select a Category>**, select **Displacement/Velocity/Acceleration**.
3. Under **<Select a Sub-Category>**, select **Linear Acceleration**.
4. Under **<Select Result Component>**, select **Y-component**.

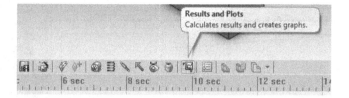

Fig. 14.23 Results and Plot option.

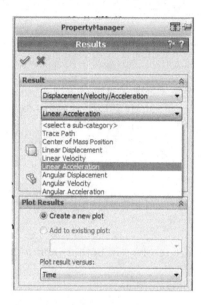

Fig. 14.24 Different results options.

The result plot for Problem 14.1 is shown in Figure 14.25 with an acceleration of 125 in/s². After 0.26 seconds, the value drops and flattens out at 0. Therefore, the solution of the simulation is taken at 0.26 seconds when the acceleration is 125 in/s².

Verification of results with analytical method

The verification of the *simulation* results with the *analytical method* is extremely important when the simulation approach is employed in problem-solving. Therefore, we compare the value of our simulation with the value obtained from the reference.

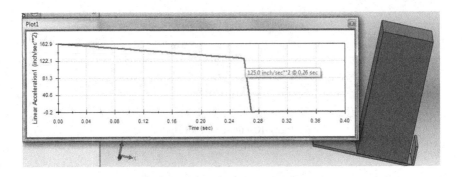

Fig. 14.25 Simulation results for Problem 14.1.

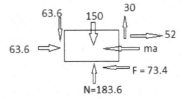

Fig. 14.26 FBD of force system.

Analytical method

1. The applied forces are resolved in both x- and y-direction. Figure 14.26 shows the FBD.

2. Applying equilibrium conditions:

$$\sum F_y = 0 : -63.6 - 150 + N + 30 = 0$$
$$N = 183.6$$
$$\therefore\ F = 0.4(183.6) = 73.44$$

$$\sum F_x = 0 : \frac{150}{32.2}a + 73.44 - 63.6 - 52 = 0$$
$$\frac{150}{32.2}a = 42.16$$
$$\therefore\ a = 9.05\,\text{ft/s}^2$$

SolidWorks simulation

The result plot for Problem 14.1 shown in Figure 14.25 gives an acceleration of $125\,\text{in/s}^2$. Converted to the same units, the acceleration is $10.416\,\text{ft/s}^2$.

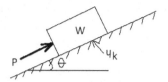

Fig. 14.27 Block on an inclined with force applied.

Our simulation result has an error of 13.87% when compared to the analytical method.

Problem 14.2

The mass of block A shown in Figure 14.27 is 30 kg (W), the applied force is 400 N (P), the coefficient of kinetic friction is 0.7 and the inclined angle is $\theta = 25°$. Determine the acceleration of the block.

SolidWorks solution

Create a model of the **Part** being studied:

1. Open a **New SolidWorks Part** document.
2. Be in **Sketch** Mode, and select **Front Plane**.
3. Sketch a **Rectangle** 0.25 by 0.25 (unit is in meters) shown in Figure 14.28.
4. Extrude 0.06 m (see Figure 14.29).
5. Apply Material: AISI 304.
6. Click **Evaluate > Mass Properties** (see Figure 14.30 for the mass properties).

Note that the dimensions have been so chosen to obtain a weight of 30 kg which is specified in the problem description. The choice of correct dimensions to obtain correct mass is an exercise for the student.

Create the **Part** model of a body on which the part being studied moves:

1. Open a **New SolidWorks Part** document.
2. Be in **Sketch** Mode, and select **Front Plane**.
3. Sketch the profile (unit is in inches) shown in Figure 14.31.
4. Extrude 0.25 m, Mid Plane (see Figure 14.32 for the Body).

The choice of material for the Body is not critical because using the **ContactManager**, we can establish the coefficient of friction between the

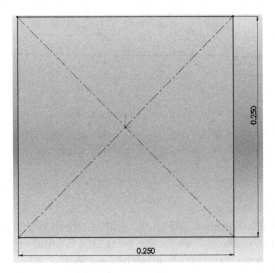

Fig. 14.28 Rectangular profile defining the block.

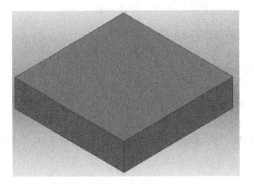

Fig. 14.29 Block.

body and the block. However, the choice of material for the Block is important in order to obtain the specified mass. Once the Block and Body are modeled, the next step is to create an assembly of both and then proceed with analysis.

Create an **Assembly** of the part and the body:

1. Open a **New SolidWorks Assembly** document.
2. Select the **Body** as the first component to **Insert** and click **OK** to accept it.

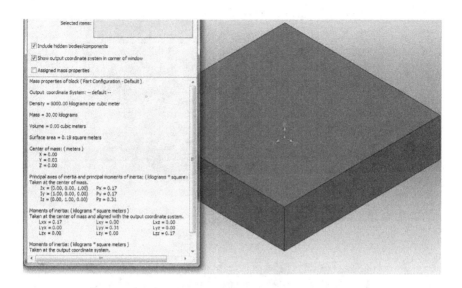

Fig. 14.30 Mass properties for the block.

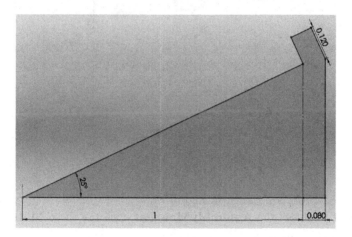

Fig. 14.31 Front profile of the body.

3. Select the **Block** as the second component to **Insert** and click **OK** (see Figure 14.33).

   Create **Mates** for the part and body (especially the **Limit Distance** mate to determine the motion constraints):

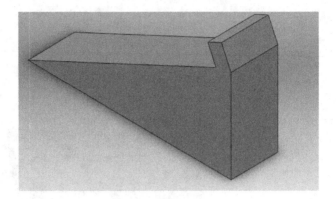

Fig. 14.32 Body.

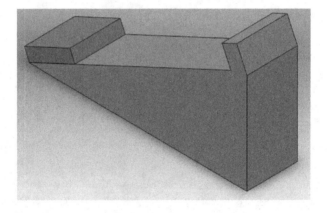

Fig. 14.33 Assembly of Body and Block.

The three **Mates** are shown in Figure 14.34.
Figure 14.35 shows the **AssemblyManager** with the three **Mates**.

Add-In the **SolidWorks Motion** tool:

1. Click **Add-In**.
2. Select the **SolidWorks Motion** tool (see Figure 14.36).

Select the **Motion Analysis** option.
Select the **Motion Analysis** option from the left side of the **MotionManager** window (see Figure 14.37).

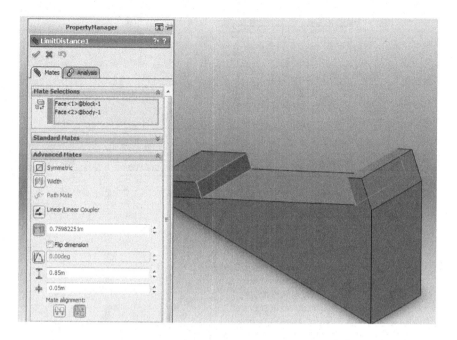

Fig. 14.34 LimitDistance1.

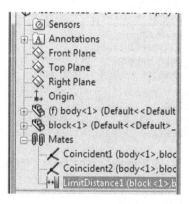

Fig. 14.35 AssemblyManager showing the three Mates.

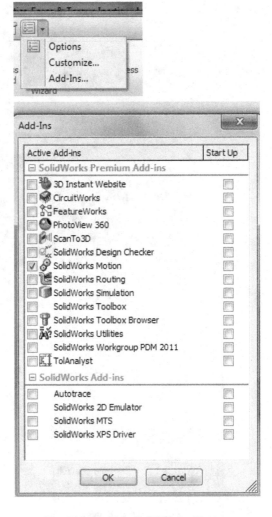

Fig. 14.36 Add-In SolidWorks Motion.

Fig. 14.37 Selecting SolidWorks Motion.

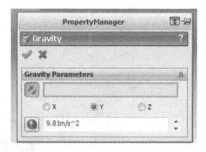

Fig. 14.38 Gravity PropertyManager.

Activate the **Gravity, Contact**, and **Force PropertyManagers**:

Gravity

1. From the **MotionManager**, click the **Gravity** tool. The **Gravity PropertyManager** automatically appears (see Figure 14.38).
2. Click the **Y**-button (Use the Gravity Parameter to choose the correct Direction; an Arrow shows up at the bottom right of the Graphics Window).
3. Click the **Check Mark** (click **OK** to finish Gravity definition).

Contact

4. From the **MotionManager**, click the **Contact** tool. The **Contact PropertyManager** automatically appears (see Figure 14.39).
5. **Uncheck** the **Material** option.
6. **Check** the **Friction** option.
7. Set **Dynamic Friction Velocity** = 0.
8. Set **Dynamic Friction Coefficient** = 0.7.
9. Set **Static Friction Velocity** = 0.
10. Set **Static Friction Coefficient** = 0.
11. In the **Elastic Properties** option, **check** the **Restitution** button.
12. Click **OK** to finish Contact definition.

Force

The applied forces are resolved in both x-direction (along the slope) and y-direction (normal to the slope). Figure 14.40 shows the FBD.

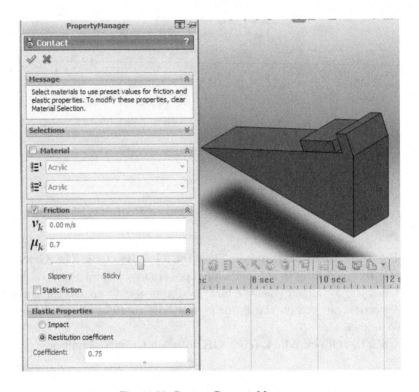

Fig. 14.39 Contact PropertyManager.

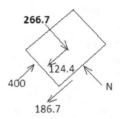

Fig. 14.40 FBD of force system.

Applying equilibrium conditions

$$\sum F_y = 0 : -266.7 + N = 0$$
$$N = 266.7$$
$$\therefore \ F = 0.7(266.7) = 186.7$$
$$\sum F_x = 400 - 124.4 - 186.7 = 88.9 \, N$$

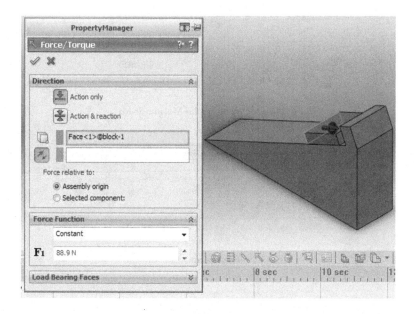

Fig. 14.41 Force/Torque PropertyManager.

From the **MotionManager**, click the **Force** tool (see Figure 14.41):

13. For the **Direction**, select the *left face of the block* (**Face <1> @block-1**) as the **Action part and point of application of action** tool (if force is applied).
14. For the **Force Direction**, click the **Reverse Direction** if the need be to obtain correct direction.
15. For the **Force Function**, select the **Constant** for the pull-down menu.
16. For the **Force Value**, type the value of **88.9 N** and click **OK** to finish Force definition.

Calculate the **Motion Study**:
Click the **Calculate** button to calculate the **Motion Study** (see Figure 14.42). Display the **Results and Plot**:

1. From the **MotionManager**, click the **Results and Plot** option (see Figure 14.43).
2. Right-click **Plot1** under **Results** folder (see Figure 14.44 for different results options).

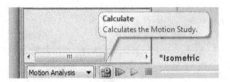

Fig. 14.42 Calculate the Motion Study.

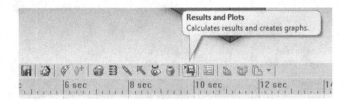

Fig. 14.43 Results and Plot option.

Fig. 14.44 Different results options.

3. Under <**Select a Category**>, select **Displacement/Velocity/Acceleration**.
4. Under <**Select a Sub-Category**>, select **Linear Acceleration**.
5. Under <**Select Result Component**>, select **Y-component**.

The result plot for Problem 2 is shown in Figure 14.25 with an acceleration of $3\,\text{m/s}^2$. After 0.46 seconds, the value drops and flattens out at 0. Therefore, the solution of the simulation is taken at 0.46 seconds when the acceleration is $3\,\text{m/s}^2$.

Verification of results with analytical method

Analytical method

1. The applied forces are resolved in both x- and y-direction. Figure 14.46 shows the FBD.
2. Applying equilibrium conditions:

$$\sum F_y = 0 : -266.7 + N = 0$$
$$N = 266.7$$
$$\therefore \ F = 0.7(266.7) = 186.7$$

$$\sum F_x = 400 - 30a - 124.4 - 186.7 = 0$$
$$a = 2.96\,\text{m/s}^2$$

SolidWorks simulation

The result plot for Problem 14.2 shown in Figure 14.45 gives an acceleration of $3\,\text{m/s}^2$. Our simulation result has and error of 1.3% when compared to the analytical method.

Problem 14.3

A 130 kg cart is accelerated horizontally by a $P = 250\,\text{N}$ force pulling at an angle of $\theta = 20°$ above horizontal (see Figure 14.47). Neglecting rolling resistance, determine the acceleration of the cart.

SolidWorks solution

Create a model of the **Part** being studied:

1. Open a **New SolidWorks Part** document.
2. Be in **Sketch** Mode, and select **Top Plane**.

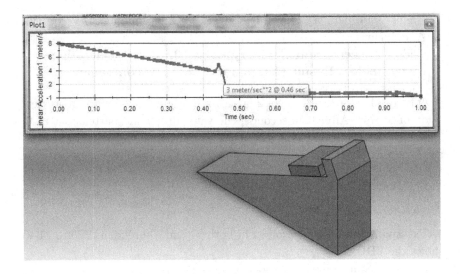

Fig. 14.45 Simulation results for Problem 14.2.

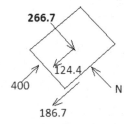

Fig. 14.46 FBD of force system.

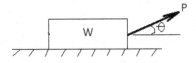

Fig. 14.47 Block on an inclined with force applied.

3. Sketch a **Rectangle** 0.25 by 0.25 (unit is in meters) shown in Figure 14.48.
4. Extrude 0.27 m (see Figure 14.49).
5. Apply Material: AISI 304.

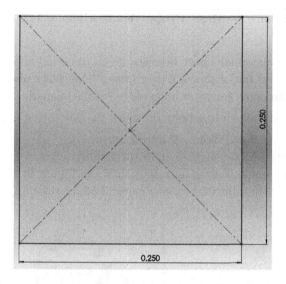

Fig. 14.48 Rectangular profile defining the block.

Fig. 14.49 Block.

6. Click **Evaluate** > **Mass Properties** (see Figure 14.50 for the mass properties).

Note that the dimensions have been so chosen to obtain a weight of 130 kg which is specified in the problem description. The choice of correct dimensions to obtain correct mass is an exercise for the student.

Create the **Part** model of a body on which the part being studied moves:

1. Open a **New SolidWorks Part** document.
2. Be in **Sketch** Mode, and select **Front Plane**.
3. Sketch the profile (unit is in inches) shown in Figure 14.51.
4. Extrude 0.25 m, Mid Plane (see Figure 14.52 for the Body).

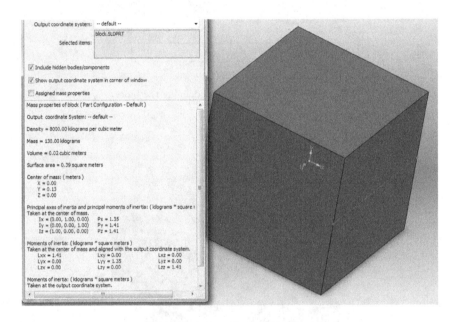

Fig. 14.50 Mass properties for the block.

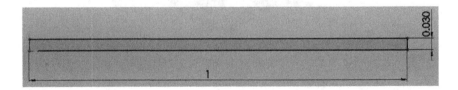

Fig. 14.51 Front profile of the body.

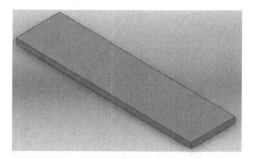

Fig. 14.52 Body.

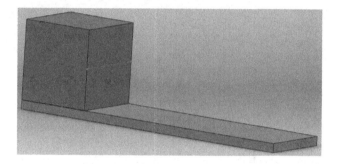

Fig. 14.53 Assembly of Body and Block.

The choice of material for the Body is not critical because using the **ContactManager**, we can establish the coefficient of friction between the body and the block. However, the choice of material for the Block is important in order to obtain the specified mass. Once the Block and Body are modeled, the next step is to create an assembly of both and then proceed with analysis.

Create an **Assembly** of the part and the body:

1. Open a **New SolidWorks Assembly** document.
2. Select the **Body** as the first component to **Insert** and click **OK** to accept it.
3. Select the **Block** as the second component to **Insert** and click **OK** (see Figure 14.53).

Create **Mates** for the part and body (especially the **Limit Distance** mate to determine the motion constraints). The three **Mates** are shown in Figure 14.54.

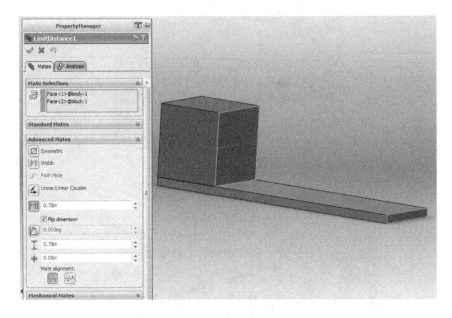

Fig. 14.54 LimitDistance1.

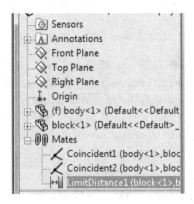

Fig. 14.55 AssemblyManager showing the three Mates.

Figure 14.55 shows the **AssemblyManager** with the three **Mates**.
Add-In the **SolidWorks Motion** tool:

1. Click **Add-In**.
2. Select the **SolidWorks Motion** tool (see Figure 14.56).

Fig. 14.56 Add-In SolidWorks Motion.

Select the **Motion Analysis** option:

Select the **Motion Analysis** option from the left side of the **MotionManager** window (see Figure 14.57).

Activate the **Gravity, Contact**, and **Force PropertyManagers**:

Gravity

1. From the **MotionManager**, click the **Gravity** tool. The **Gravity PropertyManager** automatically appears (see Figure 14.58).

Fig. 14.57 Selecting SolidWorks Motion.

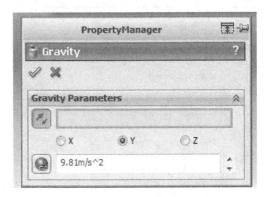

Fig. 14.58 Gravity PropertyManager.

2. Click the **Y**-button (use the Gravity Parameter to choose the correct direction; an arrow shows up at the bottom right of the Graphics Window).
3. Click the **Check Mark** (click **OK** to finish Gravity definition).

Contact

4. From the **MotionManager**, click the **Contact** tool. The **Contact PropertyManager** automatically appears (see Figure 14.59).
5. **Uncheck** the **Material** option.
6. **Check** the **Friction** option.
7. Set **Dynamic Friction Velocity** = 0.
8. Set **Dynamic Friction Coefficient** = 0.
9. Set **Static Friction Velocity** = 0.
10. Set **Static Friction Coefficient** = 0.
11. In the **Elastic Properties** option, **check** the **Restitution** button.
12. Click **OK** to finish Contact definition.

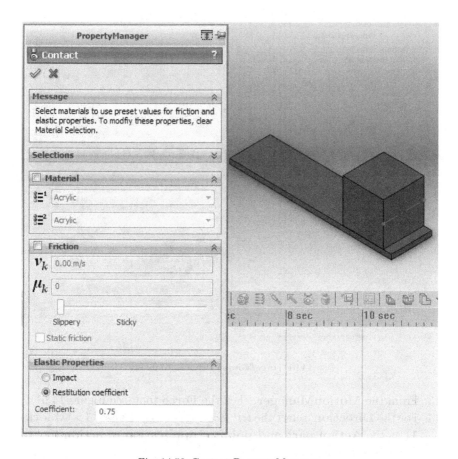

Fig. 14.59 Contact PropertyManager.

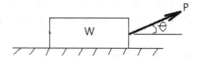

Fig. 14.60 Force system remains the same in this case.

Force

In this example, a plane PLANE1 is defined at the same angle of 20° to the horizontal as shown in Figure 14.60. No extra initial pre-processing is needed.

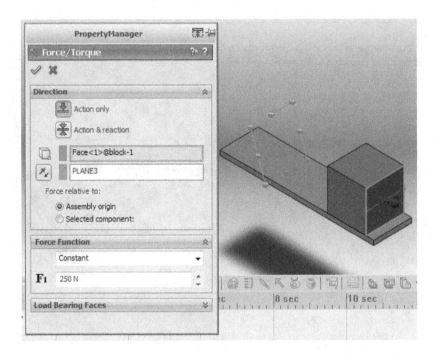

Fig. 14.61 Force/Torque PropertyManager.

13. From the **MotionManager**, click the **Force** tool (see Figure 14.61).
14. For the **Direction**, select the *left face of the block* (**Face <1> @block-1**) as the **Action part and point of application of action** tool (if force is applied).
15. For the **Force Direction**, click the **Reverse Direction** if the need be to obtain correct direction.
16. For the **Force Function**, select the **Constant** for the pull-down menu.
17. For the **Force Value**, type the value of **250 N** (note: PLANE1 is created for the loading).
18. Click **OK** to finish Force definition.

Calculate the **Motion Study**:
Click the **Calculate** button to calculate the **Motion Study** (see Figure 14.62).
Display the **Results and Plot**:

1. From the **MotionManager**, click the **Results and Plot** option (see Figure 14.63).

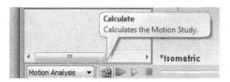

Fig. 14.62 Calculate the Motion Study.

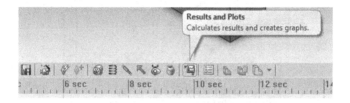

Fig. 14.63 Results and Plot option.

2. Right-click **Plot1** under **Results** folder (see Figure 14.64 for different results options).
3. Under **<Select a Category>**, select **Displacement/Velocity/Acceleration**.
4. Under **<Select a Sub-Category>**, select **Linear Acceleration**.
5. Under **<Select Result Component>**, select **Y-component**.

The result plot for Problem 3 is shown in Figure 14.65 with an acceleration of $1.8 \, \text{m/s}^2$. After 0.87 seconds, the value drops and flattens out at 0. Therefore, the solution of the simulation is taken at 0.87 seconds when the acceleration is $1.8 \, \text{m/s}^2$.

Verification of results with analytical method

Analytical method

1. The applied forces are resolved in both x- and y-direction. Figure 14.66 shows the FBD.
2. Applying equilibrium conditions

$$\sum F_x = 0 : \quad 130a - 250 \, Cos \, 20 = 0$$
$$\therefore \, a = 1.8 \, \text{m/s}^2$$

Fig. 14.64 Different results options.

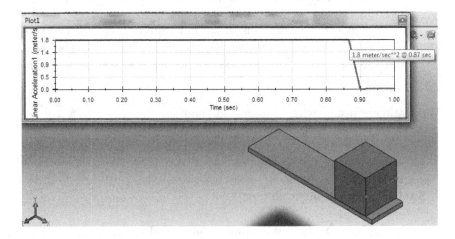

Fig. 14.65 Simulation results for Problem 14.3.

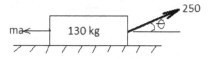

Fig. 14.66 FBD of force system.

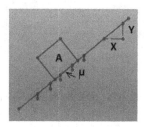

Fig. 14.67 Block A on an incline.

SolidWorks simulation

The result plot for Problem 14.3 shown in Figure 14.65 gives an acceleration of $1.8\,\mathrm{m/s^2}$. Our simulation result has no error when compared to the analytical method.

Problem 14.4

Determine the acceleration of block A shown in Figure 14.67, mass $= 45\,\mathrm{kg}$, down the slope, for y $= 5$, x $= 12$, $\mu = 0.2$.

SolidWorks solution

Create a model of the **Part** being studied:

1. Open a **New SolidWorks Part** document.
2. Be in **Sketch** Mode, and select **Top Plane**.
3. Sketch a **Rectangle** 0.25 by 0.25 (unit is in inches) shown in Figure 14.68.
4. Extrude 0.09 m (see Figure 14.69).
5. Apply Material: AISI 304.
6. Click **Evaluate > Mass Properties** (see Figure 14.70 for the mass properties).

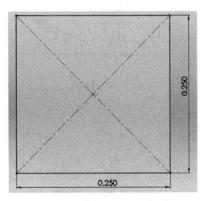

Fig. 14.68 Rectangular profile defining the block.

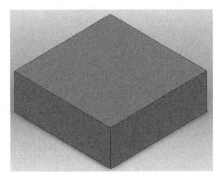

Fig. 14.69 Block.

Note that the dimensions have been so chosen to obtain a mass of 45 kg which is specified in the problem description. The choice of correct dimensions to obtain correct mass is an exercise for the student.

Create the **Part** model of a body on which the part being studied moves:

1. Open a **New SolidWorks Part** document.
2. Be in **Sketch** Mode, and select **Front Plane**.
3. Sketch the profile (unit is in inches) shown in Figure 14.71.
4. Extrude 0.09 m (see Figure 14.72 for the Body).

The choice of material for the Body is not critical because using the **ContactManager**, we can establish the coefficient of friction between the

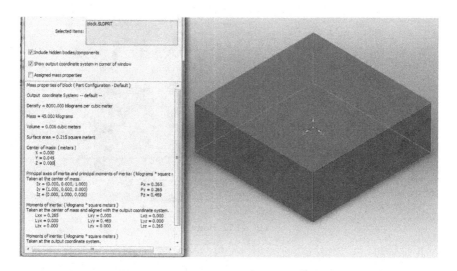

Fig. 14.70 Mass properties for the block.

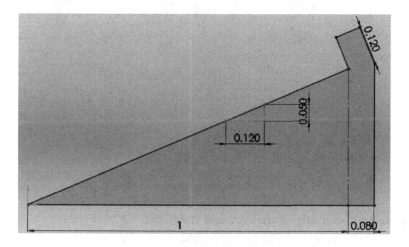

Fig. 14.71 Front profile of the body.

body and the block. However, the choice of material for the Block is important in order to obtain the specified mass. Once the Block and Body are modeled, the next step is to create an assembly of both and then proceed with analysis.

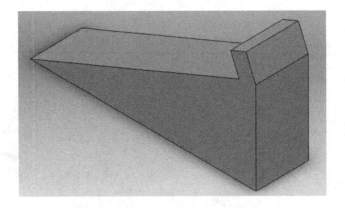

Fig. 14.72 Body.

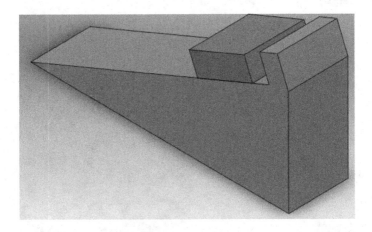

Fig. 14.73 Assembly of Body and Block.

Create an **Assembly** of the part and the body:

1. Open a **New SolidWorks Assembly** document.
2. Select the **Body** as the first component to **Insert** and click **OK** to accept it.
3. Select the **Block** as the second component to **Insert** and click **OK** (see Figure 14.73).

Create **Mates** for the part and body (especially the **Limit Distance** mate to determine the motion constraints):

The three **Mates** are shown in Figure 14.74. Figure 14.75 shows the **AssemblyManager** with the three **Mates**.

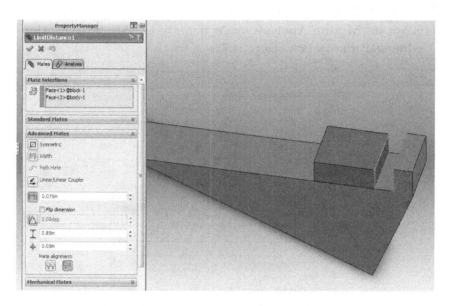

Fig. 14.74 LimitDistance1.

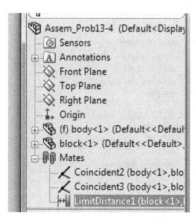

Fig. 14.75 AssemblyManager showing the three Mates.

Add-In the **SolidWorks Motion** tool:

1. Click **Add-In**.
2. Select the **SolidWorks Motion** tool (see Figure 14.76).

Select the **Motion Analysis** option:

1. Select the **Motion Analysis** option from the left side of the **Motion-Manager** window (see Figure 14.77).

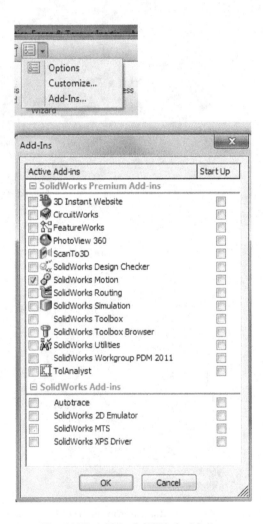

Fig. 14.76 Add-In SolidWorks Motion.

Fig. 14.77 Selecting SolidWorks Motion.

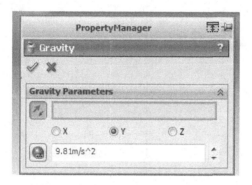

Fig. 14.78 Gravity PropertyManager.

Activate the **Gravity**, **Contact**, and **Force PropertyManagers**:
Gravity

1. From the **MotionManager**, click the **Gravity** tool. The **Gravity PropertyManager** automatically appears (see Figure 14.78).
2. Click the **Y**-button (use the Gravity Parameter to choose the correct direction; an arrow shows up at the bottom right of the Graphics Window).
3. Click the **Check Mark (OK)**.

Contact

4. From the **MotionManager**, click the **Contact** tool. The **Contact PropertyManager** automatically appears (see Figure 14.79).
5. **Uncheck** the **Material** option.
6. **Check** the **Friction** option.
7. Set **Dynamic Friction Velocity** = 0.
8. Set **Dynamic Friction Coefficient** = 0.
9. Set **Static Friction Velocity** = 0.
10. Set **Static Friction Coefficient** = 0.2.
11. In the **Elastic Properties** option, **check** the **Restitution** button.

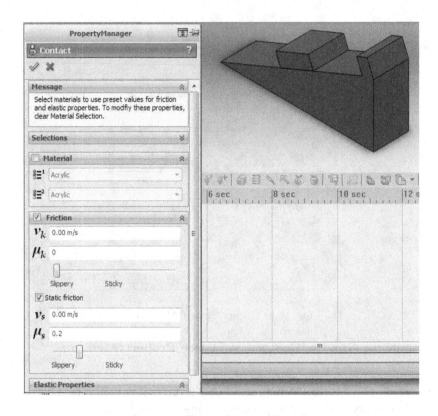

Fig. 14.79 Contact PropertyManager.

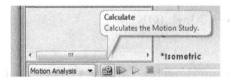

Fig. 14.80 Calculate the Motion Study.

Force

In this example, no extra initial pre-processing is needed.

Calculate the **Motion Study**:

Click the **Calculate** button to calculate the **Motion Study** (see Figure 14.80).

Display the **Results and Plot**:
From the **MotionManager**, click the **Results and Plot** option (see
Figure 14.81).

1. Right-click **Plot1** under **Results** folder (see Figure 14.82 for different
   results options).

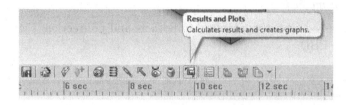

Fig. 14.81 Results and Plot option.

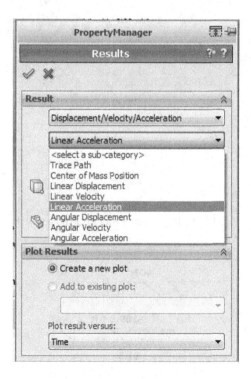

Fig. 14.82 Different results options.

2. Under **<Select a Category>**, select **Displacement/Velocity/ Acceleration**.
3. Under **<Select a Sub-Category>**, select **Linear Acceleration**.
4. Under **<Select Result Component>**, select **Y-component**.

The result plot for Problem 14.4 is shown in Figure 14.83 with an acceleration of $3.48\,\text{m/s}^2$.

Analytical method

1. The applied forces are resolved in both x- and y-direction. Figure 14.84 shows the FBD.

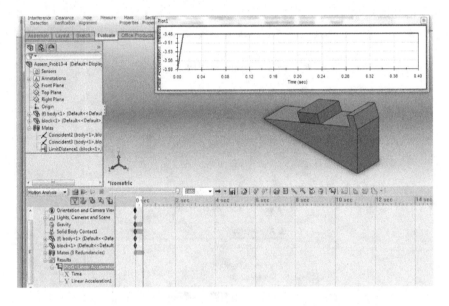

Fig. 14.83 Simulation results for Problem 14.4.

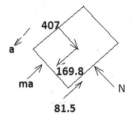

Fig. 14.84 FBD of force system.

2. Applying equilibrium conditions.

$$\sum F_y = 0: \ -407 + N = 0$$
$$N = 407$$
$$\therefore \ F = 0.2(407) = 81.5$$
$$\sum F_x = 45a + 81.5 - 170 = 0$$
$$a = 1.97 \, \text{m/s}^2$$

SolidWorks simulation

The result plot for Problem 14.4 shown in Figure 14.84 gives an acceleration of $3.48 \, \text{m/s}^2$. Our simulation result has excess error compared to the analytical method.

Problem 14.5

Determine the acceleration of block A shown in Figure 14.85 for the block weights $W_1 = 80 \, \text{lb}$, $W_2 = 20 \, \text{lb}$ and $P = 30.6 \, \text{lb}$. Assume $\mu_k = 0.4$.

SolidWorks solution

Create a model of the **Part** being studied:

1. Open a **New SolidWorks Part** document.
2. Be in **Sketch** Mode, and select **Front Plane**.
3. Sketch a **Rectangle** 6 by 15 (unit is in inches) shown in Figure 14.86.
4. Extrude 3 in (see Figure 14.87).
5. Apply Material: AISI 304.

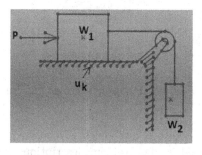

Fig. 14.85 Block, pulley, body system.

Fig. 14.86 Rectangular profile defining the block.

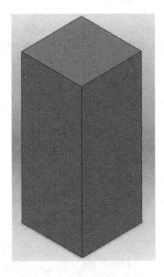

Fig. 14.87 Block.

6. Click **Evaluate** > **Mass Properties** (see Figure 14.88 for the mass properties).

Note that the dimensions have been so chosen to obtain a weight of 80 lb which is specified in the problem description. The choice of correct dimensions to obtain correct mass is an exercise for the student.

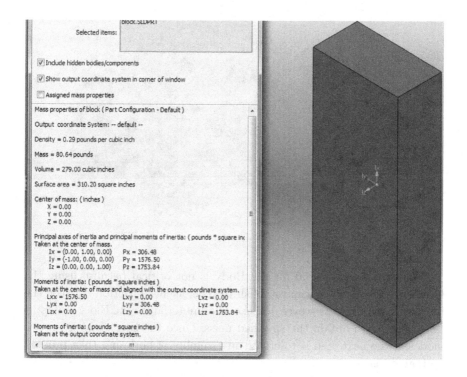

Fig. 14.88 Mass properties for the block.

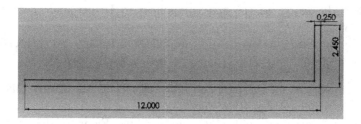

Fig. 14.89 Front profile of the body.

Create the **Part** model of a body on which the part being studied moves:

1. Open a **New SolidWorks Part** document.
2. Be in **Sketch** Mode, and select **Front Plane**.
3. Sketch the profile (unit is in inches) shown in Figure 14.89.
4. Extrude 6 in, Mid Plane (see Figure 14.90 for the Body).

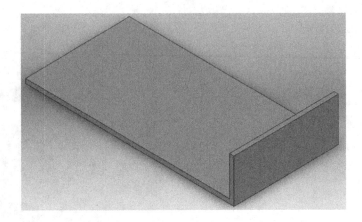

Fig. 14.90 Body.

The choice of material for the Body is not critical because using the **ContactManager**, we can establish the coefficient of friction between the body and the block. However, the choice of material for the Block is important in order to obtain the specified mass. Once the Block and Body are modeled, the next step is to create an assembly of both and then proceed with analysis.

Create an **Assembly** of the part and the body:

1. Open a **New SolidWorks Assembly** document.
2. Select the **Body** as the first component to **Insert** and click **OK** to accept it.
3. Select the **Block** as the second component to **Insert** and click **OK** (see Figure 14.91).

Create **Mates** for the part and body (especially the **Limit Distance** mate to determine the motion constraints):
The three **Mates** are shown in Figure 14.92. Figure 14.93 shows the **AssemblyManager** with the three **Mates**.

Add-In the **SolidWorks Motion** tool:

1. Click **Add-In**.
2. Select the **SolidWorks Motion** tool (see Figure 14.94).

Select the **Motion Analysis** option:
Select the **Motion Analysis** option from the left side of the **MotionManager** window (see Figure 14.95).

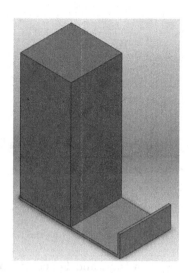

Fig. 14.91 Assembly of Body and Block.

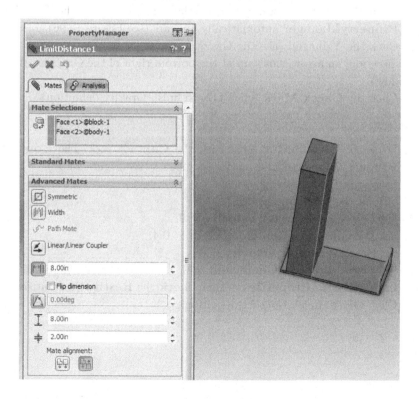

Fig. 14.92 LimitDistance1.

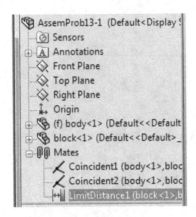

Fig. 14.93 AssemblyManager showing the three Mates.

Activate the **Gravity**, **Contact**, and **Force PropertyManagers**:
Gravity

1. From the **MotionManager**, click the **Gravity** tool. The **Gravity PropertyManager** automatically appears (see Figure 14.96).
2. Click the **Y**-button (use the Gravity Parameter to choose the correct direction; an arrow shows up at the bottom right of the Graphics Window).
3. Click the **Check Mark** (click **OK** to finish Gravity definition).

Contact

4. From the **MotionManager**, click the **Contact** tool. The **Contact PropertyManager** automatically appears (see Figure 14.97).
5. **Uncheck** the **Material** option.
6. **Check** the **Friction** option.
7. Set **Dynamic Friction Velocity** = 0.
8. Set **Dynamic Friction Coefficient** = 0.4.
9. Set **Static Friction Velocity** = 0.
10. Set **Static Friction Coefficient** = 0.
11. In the **Elastic Properties** option, **check** the **Restitution** button.
12. Click **OK** to finish Contact definition.

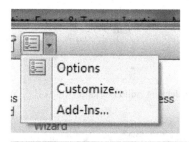

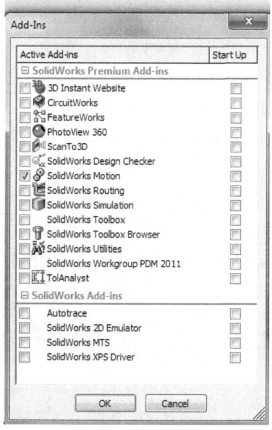

Fig. 14.94 Add-In SolidWorks Motion.

Fig. 14.95 Selecting SolidWorks Motion.

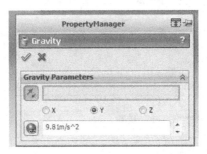

Fig. 14.96 Gravity PropertyManager.

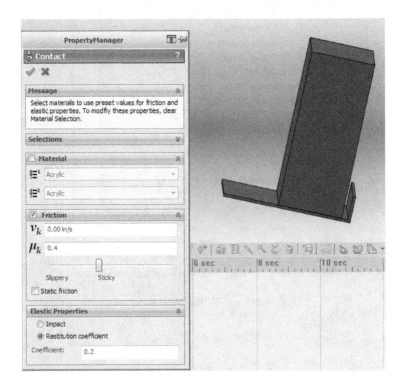

Fig. 14.97 Contact PropertyManager.

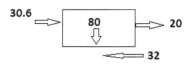

Fig. 14.98 FBD of force system.

Force

The applied forces are resolved in both x and y-direction. Figure 14.98 shows the FBD.

Applying equilibrium conditions

$$\sum F_y = 0 : -80\_N = 0$$
$$N = 80$$
$$\therefore \ F = 0.4(80) = 32$$

$$\sum F_x = 0 : 20 + 30.6 - 32 = 18.6 \, \text{lb}$$

From the **MotionManager**, click the **Force** tool (see Figure 14.99).

13. For the **Direction**, select the *left face of the block* (**Face <1> @block-1**) as the **Action part and point of application of action** tool (if force is applied).
14. For the **Force Direction**, click the **Reverse Direction** if the need be to obtain correct direction.
15. For the **Force Function**, select the **Constant** for the pull-down menu.
16. For the **Force Value**, type the value of **18.6 lbf** and click **OK** to finish Force definition.

Calculate the **Motion Study**:

Click the **Calculate** button to calculate the **Motion Study** (see Figure 14.100).

Display the **Results and Plot**:

From the **MotionManager**, click the **Results and Plot** option (see Figure 14.101).

1. Right-click **Plot1** under **Results** folder (see Figure 14.102 for different results options).
2. Under **<Select a Category>**, select **Displacement/Velocity/Acceleration**.
3. Under **<Select a Sub-Category>**, select **Linear Acceleration**.
4. Under **<Select Result Component>**, select **Y-component**.

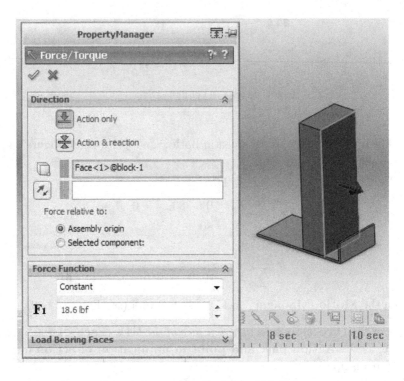

Fig. 14.99 Force/Torque PropertyManager.

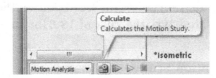

Fig. 14.100 Calculate the Motion Study.

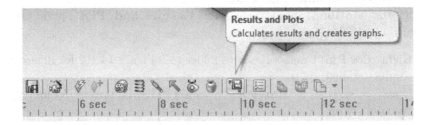

Fig. 14.101 Results and Plot option.

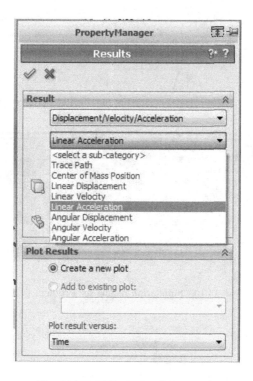

Fig. 14.102 Different results options.

The result plot for Problem 14.5 is shown in Figure 14.103 with an acceleration of $96\,\text{in/s}^2$. After 0.33 seconds, the value drops and flattens out at 0. Therefore, the solution of the simulation is taken at 0.33 seconds when the acceleration is $96\,\text{in/s}^2$.

Verification of results with analytical method

Analytical method

1. The applied forces are resolved in both x- and y-direction. Figure 14.104 shows the FBD.

2. Applying equilibrium conditions
For $W_2 = 20\,\text{lb}$:

$$\sum F_y = 0 : T + \frac{20}{32}a - 20 = 0$$

$$\therefore T = 20 - 0.6211a$$

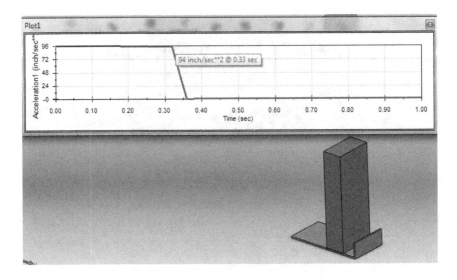

Fig. 14.103 Simulation results for Problem 14.5

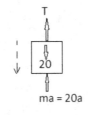

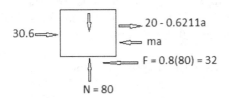

Fig. 14.104 FBD of force system.

For $W_1 = 80\,0$lb:

$$\sum F_y = 0 : -80 + N = 0$$
$$N = 80$$
$$\therefore \ F = 0.4(80) = 32$$

$$\sum F_x = 0 : 30.6 - \frac{80}{32.2}a - 32 + (20 - 0.6211a) = 0$$
$$(2.484 + 0.6211)a = 30.6 - 32 + 20 = 18.6$$
$$3.1051a = 18.6$$
$$\therefore \ a = 6\,\text{ft/s}^2$$

SolidWorks simulation

The result plot for Problem 14.5 shown in Figure 14.103 gives an acceleration of 94 in/s². Converted to the same units, the acceleration is 7.83 ft/s². Our simulation result deviates a bit when compared to the analytical method, but it is quite acceptable.

Kinetics: Plane Motion

So far, we have dealt with kinetics problems that are linear in nature. We now deal with kinetics problems that are in plane motion.

Problem 14.6

In Figure 14.105, cylinder A has a mass of 70 lb, and mass B is 20 lb. The coefficient of friction between the cylinder and the floor is 0.70. The diameter of the cylinder is D = 24 in and the dimension y = 8 in. Determine (system is initially at rest):

(a) The acceleration at the center of the cylinder A.
(b) The acceleration of mass B.

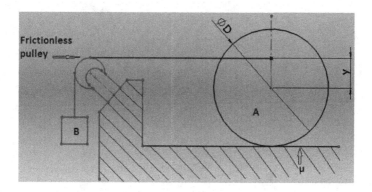

Fig. 14.105 Plane motion system.

(c) Frictional force acting on cylinder A.

(d) Linear distance A travels in 3 seconds.

Note: All dimensions are in the US customary system.

SolidWorks solution

Create a model of the **Part** being studied:

1. Open a **New SolidWorks Part** document.
2. Be in **Sketch** Mode, and select **Front Plane**.
3. Sketch a **Circle** 24 in diameter shown in Figure 14.106.
4. Extrude 0.546 in (see Figure 14.107). Note that three additional bosses and a point (Point1) are added; these features will be used in defining forces, etc.
5. Apply Material: AISI 304.
6. Click **Evaluate** > **Mass Properties** (see Figure 14.108 for the mass properties).

Density $= 0.28 \, \text{lb/in}^3$

Mass $= 70.74 \, \text{lb}$

Volume $= 248.81 \, \text{in}^3$

Surface area $= 946.25 \, \text{in}^2$

Center of mass (inches): X $= 0.00$; Y $= 0.00$; Z $= 0.00$

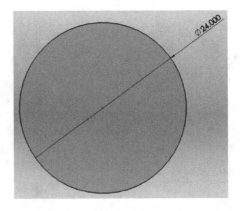

Fig. 14.106 Circular profile defining the cylinder.

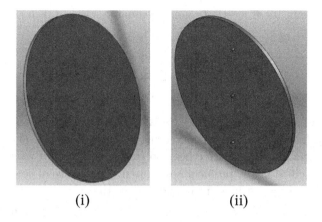

(i)   (ii)

Fig. 14.107 Cylinder.

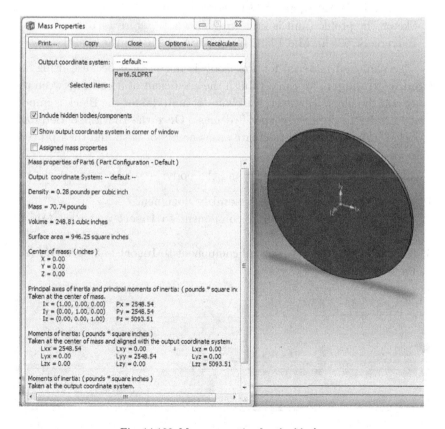

Fig. 14.108 Mass properties for the block.

Principal moments of inertia (pounds * square inches), taken at the center of mass:

$$Px = 2548.54$$
$$Py = 2548.54$$
$$Pz = 5093.51$$

Moment of inertia $= 5093.51/(32.2 \times 12 \times 12) = 1.098$ lb-ft-s^2.

This values agrees with theory using $1/2\,mr^2 = 1/2\frac{70}{32.2}(1)^2 = 1.087$. Note that the dimensions have been so chosen to obtain a weight of 70 lb which is specified in the problem description. The choice of correct dimensions to obtain correct mass is an exercise for the student.

Create the **Part** model of a body on which the part being studied moves:

1. Open a **New SolidWorks Part** document.
2. Be in **Sketch** Mode, and select **Front Plane**.
3. Sketch the profile (unit is in inches) shown in Figure 14.109.
4. Extrude 8.7625 in, Mid Plane (see Figure 14.110 for the Body).

The choice of material for the Body is not critical because using the **ContactManager**, we can establish the coefficient of friction between the body and the block. However, the choice of material for the Block is important in order to obtain the specified mass. Once the Block and Body are modeled, the next step is to create an assembly of both and then proceed with analysis.

Create an **Assembly** of the part and the body:

1. Open a **New SolidWorks Assembly** document.
2. Select the **Body** as the first component to **Insert** and click **OK** to accept it.
3. Select the **Block** as the second component to **Insert** and click **OK** (see Figure 14.111).

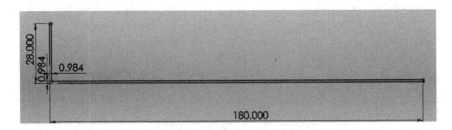

Fig. 14.109 Front profile of the body.

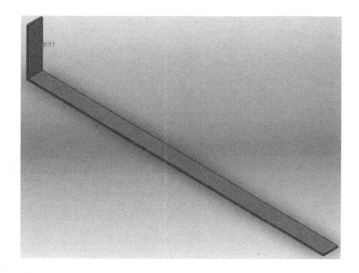

Fig. 14.110 Body.

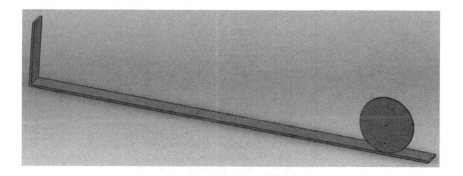

Fig. 14.111 Assembly of Body and Block.

Create **Mates** for the part and body (especially the **Limit Distance** mate to determine the motion constraints):

The three **Mates** are shown in Figure 14.112.

Figure 14.113 shows the **AssemblyManager** with the three **Mates**.

Add-In the **SolidWorks Motion** tool:

1. Click **Add-In**.
2. Select the **SolidWorks Motion** tool (see Figure 14.114).

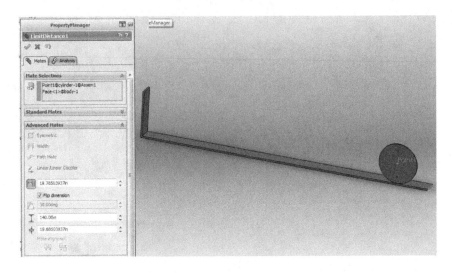

Fig. 14.112 LimitDistance1.

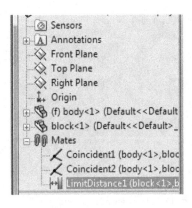

Fig. 14.113 AssemblyManager showing the three Mates.

Select the **Motion Analysis** option:

Select the **Motion Analysis** option from the left side of the **MotionManager** window (see Figure 14.115).

Activate the **Gravity, Contact,** and **Force PropertyManagers**:

Gravity

1. From the **MotionManager**, click the **Gravity** tool. The **Gravity PropertyManager** automatically appears (see Figure 14.116).

Fig. 14.114 Add-In SolidWorks Motion.

Fig. 14.115 Selecting SolidWorks Motion.

2. Click the **Y**-button (use the Gravity Parameter to choose the correct direction; an arrow shows up at the bottom right of the Graphics Window).
3. Click the **Check Mark** (click **OK** to finish Gravity definition).

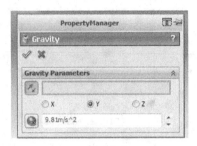

Fig. 14.116 Gravity PropertyManager.

Contact

4. From the **MotionManager**, click the **Contact** tool. The **Contact PropertyManager** automatically appears (see Figure 14.117).
5. **Uncheck** the **Material** option.
6. **Check** the **Friction** option.
7. Set **Dynamic Friction Velocity** = 0.
8. Set **Dynamic Friction Coefficient** = 0.4.
9. Set **Static Friction Velocity** = 0.
10. Set **Static Friction Coefficient** = 0.7.
11. In the **Elastic Properties** option, **check** the **Restitution** button.
12. Click **OK** to finish Contact definition.

Force
No pre-processing is applied to this class of problem.
From the **MotionManager**, click the **Force** tool (see Figure 14.118).

13. For the **Direction**, select the *front face of the topmost boss* (**Face <1> @cylinder-1**) as the **Action part and point of application of action** tool.
14. For the **Force Direction**, click the *inner face of the body* and also click **Reverse Direction** if the need be to obtain correct direction.
15. For the **Force Function**, select the **Constant** for the pull-down menu.
16. For the **Force Value**, type the value of **20 lbf** and click **OK** to finish Force definition.

Calculate the **Motion Study**:
Click the **Calculate** button to calculate the **Motion Study** (see Figure 14.119).

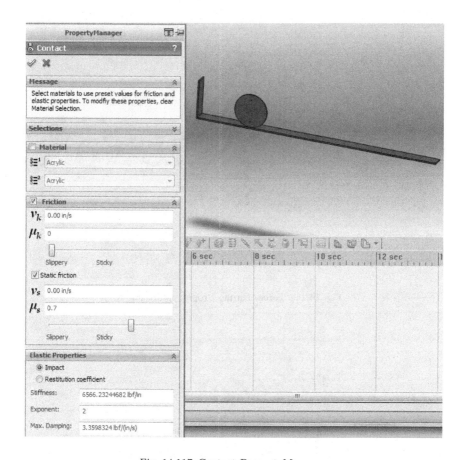

Fig. 14.117 Contact PropertyManager.

Display the **Results and Plot**:
From the **MotionManager**, click the **Results and Plot** option (see Figure 14.120).

1. Right-click **Plot1** under **Results** folder (see Figure 14.121 for different results options).
2. Under <**Select a Category**>, select **Displacement/Velocity/ Acceleration**.
3. Under <**Select a Sub-Category**>, select **Linear Acceleration**.
4. Under <**Select Result Component**>, select **Y-component**.

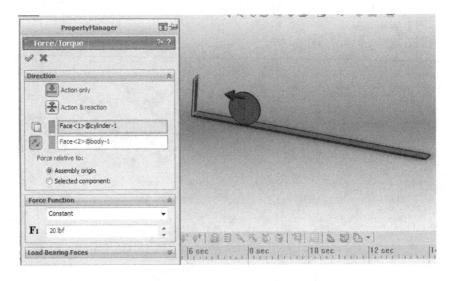

Fig. 14.118 Force/Torque PropertyManager.

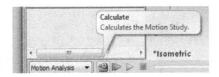

Fig. 14.119 Calculate the Motion Study.

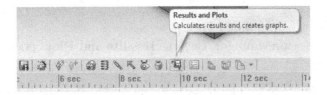

Fig. 14.120 Results and Plot option.

The result plot for Problem 14.6 is shown in Figure 14.122 with an acceleration of $115 \, \text{in/s}^2$. After 1.34 seconds, the value drops and flattens out at 0. Therefore, the solution of the simulation is taken at 1.34 seconds when the acceleration is $115 \, \text{in/s}^2$.

Fig. 14.121 Different results options.

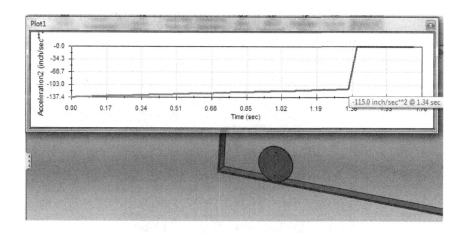

Fig. 14.122 Simulation results for Problem 14.6

Verifications of results with analytical method

Analytical method

1. Motion Analysis — when the center of A travels with an acceleration a, mass B travels with an acceleration $(1 + 0.667)a = 1.667a$ (see Figure 14.123). When mass B drops, mass A rotates counter-clockwise.

From the triangles formed, if mass B travels 1 in, the center of the cylinder will travel 0.6 in. Therefore, we have the relationships:

$$S_c = 0.6S_B; \quad V_c = 0.6V_B$$

Conversely, $S_B = 1.667S_C; V_B = 1.667V_C$.

2. Figure 14.124 shows the FBDs.

3. Applying equilibrium conditions:
For Cylinder:

$$\sum M_o = 0 : T(1.667) + I_C\alpha - (m_A\alpha)r = 0$$

$$T(1.667) = 1.087\left(\frac{r}{1}\right) + \left(\frac{70}{32.2}\right)(a)(1)$$

$$T = \left(\frac{1.087}{1.667}\right)\left(\frac{r}{1}\right) + \left(\frac{70}{1.667 \times 32.2}\right)(a) = 0.652a + 1.304a$$

$$\therefore T = 1.956a \tag{14.1}$$

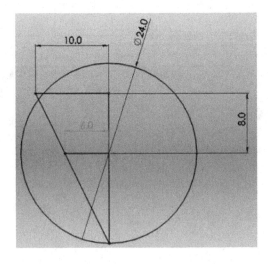

Fig. 14.123 Rolling wheel relations.

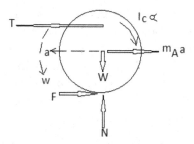

Fig. 14.124 FBD of force system.

For mass B:

$$\sum F_y = 0 : 20 - m_B a_B - T = 0$$

$$T = 20 - \left(\frac{20}{32.2}\right)(1.667a) = 20 - 1.0354a \qquad (14.2)$$

Using equations (14.1) and (14.2): $1.956a = 20 - 1.0354a$, leading to:

$$\sum F_x = 0 : 30.6 - a - 32 + (20 - 0.6211a) = 0$$

$$(2.484 + 0.6211)a = 30.6 - 32 + 20 = 18.6$$

$$a = \frac{20}{2.9912}$$

$$\therefore \ a = 6.686 \, \text{ft/s}^2$$

The acceleration of B, $a_B = 1.667a = 11.146 \, \text{ft/s}^2$. Frictional force acting on cylinder A:

$$\sum F_x = 0 : m_A \cdot a - T + F = 0$$

$$F = T - m_A \cdot a = 1.956a - \frac{70}{32.2} \cdot a = 1.956a - 2.17a = -0.214a$$

$$\therefore \ F = -1.43 \, \text{lb}$$

The displacement A travels in 3 seconds:

$$s = v_o t + \frac{1}{2}at^t = 0 + \frac{1}{2}(6.686)(3)^2 = 30.087 \, \text{ft}$$

SolidWorks simulation

The result plot for Problem 14.6 shown in Figure 14.122 gives an acceleration of $115 \, \text{in/s}^2$. Converted to the same units, the acceleration is $9.58 \, \text{ft/s}^2$. Our simulation result deviates a bit when compared to the analytical method, but it is quite acceptable. As can be observed, SolidWorks simulation is a very effective tool for analyzing dynamics of bodies with plane motion. The simulation can adapt to changes made to the shape of object being investigated with very little penalty in computation time. The analytical method is long and prone to error by the user when compared to using SolidWorks simulation.

Problem 14.7

In Figure 14.125, cylinder A has a mass of 140 lb, and mass B is 25 lb. The coefficient of friction between the cylinder and the floor is 0.70. The diameter of the cylinder is D = 48 in and the dimension y = 16 in. Determine (system is initially at rest):

(a) The acceleration at the center of the cylinder A.
(b) The acceleration of mass B.
(c) Frictional force acting on cylinder A.
(d) Linear distance A travels in 3 seconds.

Note: All dimensions are in US customary system.

SolidWorks solution

Create a model of the **Part** being studied:

1. Open a **New SolidWorks Part** document.
2. Be in **Sketch** Mode, and select **Front Plane**.

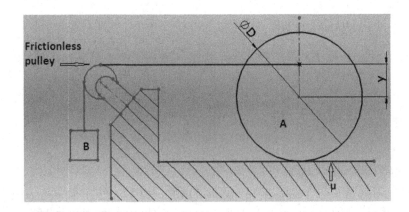

Fig. 14.125 Dynamics system.

3. Sketch a **Circle** 48 in diameter shown in Figure 14.126.
4. Extrude 0.273 in (see Figure 14.127). Note that three additional bosses and a point (Point1) are added; these features will be used in defining forces, etc.
5. Apply Material: AISI 304.

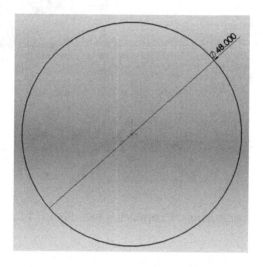

Fig. 14.126 Circular profile defining the cylinder.

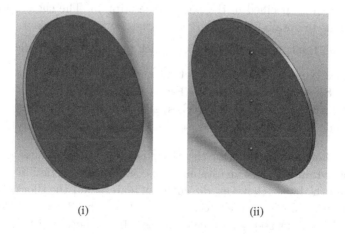

(i)   (ii)

Fig. 14.127 Cylinder.

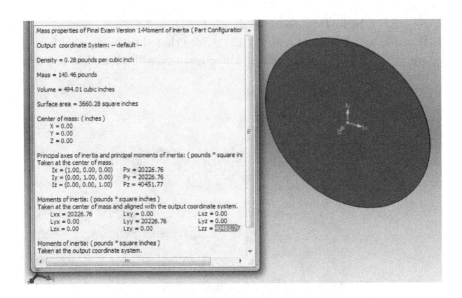

Fig. 14.128 Mass properties for the cylinder.

6. Click **Evaluate > Mass Properties** (see Figure 14.128 for the mass properties).

Moment of inertia $= 40451.77/(32.2 \times 12 \times 12) = 8.72\,\text{lb-ft-s}^2$. This values agrees with theory using $1/2\,\text{mr}^2 = 1/2\frac{140}{32.2}(2)^2 = 8.7$

Note that the dimensions have been so chosen to obtain a weight of 140 lb which is specified in the problem description. The choice of correct dimensions to obtain correct mass is an exercise for the student.

Create the **Part** model of a body on which the part being studied moves:

1. Open a **New SolidWorks Part** document.
2. Be in **Sketch** Mode, and select **Front Plane**.
3. Sketch the profile (unit is in inches) shown in Figure 14.129.
4. Extrude 8.7625 in, Mid Plane (see Figure 14.130 for the Body).

The choice of material for the Body is not critical because using the **ContactManager,** we can establish the coefficient of friction between the body and the block. However, the choice of material for the Block is important in order to obtain the specified mass. Once the Block and Body are modeled, the next step is to create an assembly of both and then proceed with analysis.

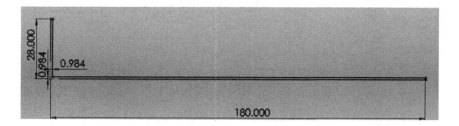

Fig. 14.129 Front profile of the body.

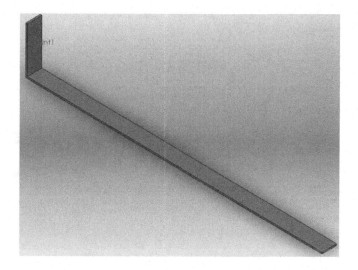

Fig. 14.130 Body.

Create an **Assembly** of the part and the body:

1. Open a **New SolidWorks Assembly** document.
2. Select the **Body** as the first component to **Insert** and click **OK** to accept it.
3. Select the **Block** as the second component to **Insert** and click **OK** (see Figure 14.131).

Create **Mates** for the part and body (especially the **Limit Distance** mate to determine the motion constraints):

The three **Mates** are shown in Figure 14.132. Figure 14.133 shows the **AssemblyManager** with the three **Mates**.

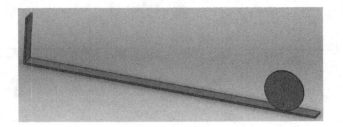

Fig. 14.131 Assembly of Body and Block.

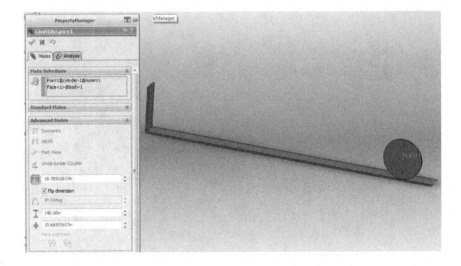

Fig. 14.132 LimitDistance1.

Add-In the **SolidWorks Motion** tool:

1. Click **Add-In**.
2. Select the **SolidWorks Motion** tool (see Figure 14.134).

Select the **Motion Analysis** option:
Select the **Motion Analysis** option from the left side of the **MotionManager** window (see Figure 14.135).

Activate the **Gravity, Contact**, and **Force PropertyManagers**:
Gravity

1. From the **MotionManager**, click the **Gravity** tool. The **Gravity PropertyManager** automatically appears (see Figure 14.136).

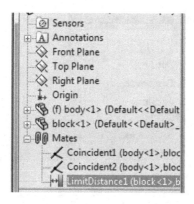

Fig. 14.133 AssemblyManager showing the three Mates.

2. Click the **Y**-button (use the Gravity Parameter to choose the correct direction; an arrow shows up at the bottom right of the Graphics Window).
3. Click the **Check Mark** (click **OK** to finish Gravity definition).

Contact

4. From the **MotionManager**, click the **Contact** tool. The **Contact PropertyManager** automatically appears (see Figure 14.137).
5. **Uncheck** the **Material** option.
6. **Check** the **Friction** option.
7. Set **Dynamic Friction Velocity** = 0.
8. Set **Dynamic Friction Coefficient** = 0.4.
9. Set **Static Friction Velocity** = 0.
10. Set **Static Friction Coefficient** = 0.7.
11. In the **Elastic Properties** option, **check** the **Restitution** button.
12. Click **OK** to finish Contact definition.

Force

No pre-processing is applied to this class of problem.

From the **MotionManager**, click the **Force** tool (see Figure 14.138).

13. For the **Direction**, select the *front face of the topmost boss* (**Face <1> @cylinder-1**) as the **Action part and point of application of action** tool.
14. For the **Force Direction**, click the *inner face of the body* and also click **Reverse Direction** if the need be to obtain correct direction.

Fig. 14.134 Add-In SolidWorks Motion.

Fig. 14.135 Selecting SolidWorks Motion.

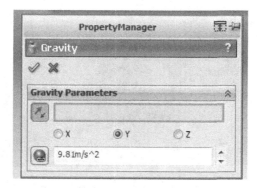

Fig. 14.136 Gravity PropertyManager.

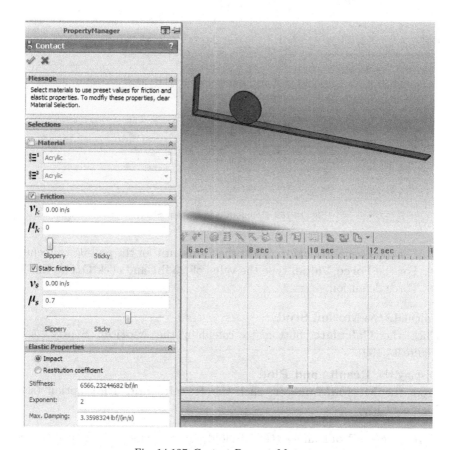

Fig. 14.137 Contact PropertyManager.

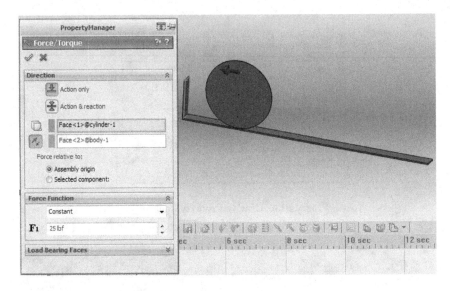

Fig. 14.138 Force/Torque PropertyManager.

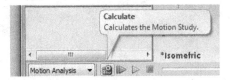

Fig. 14.139 Calculate the Motion Study.

15. For the **Force Function**, select the **Constant** for the pull-down menu.
16. For the **Force Value**, type the value of **25 lbf** and click **OK** to finish Force definition.

Calculate the **Motion Study**:
Click the **Calculate** button to calculate the **Motion Study** (see Figure 14.139).

Display the **Results and Plot**:
From the **MotionManager**, click the **Results and Plot** option (see Figure 14.140).

1. Right-click **Plot1** under **Results** folder (see Figure 14.141 for different results options).

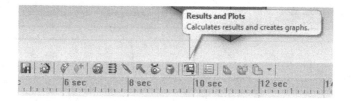

Fig. 14.140 Results and Plot option.

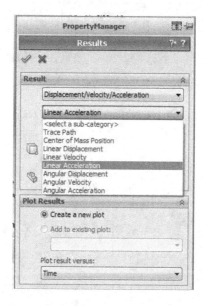

Fig. 14.141 Different results options.

2. Under <Select a Category>, select **Displacement/Velocity/ Acceleration**.
3. Under <**Select a Sub-Category**>, select **Linear Acceleration**.
4. Under <**Select Result Component**>, select **Y-component**.

The result plot for Problem 14.7 is shown in Figure 14.142 with an acceleration of 68.7 in/s². After 1.83 seconds, the value drops and flattens out at 0. Therefore, the solution of the simulation is taken at 1.83 seconds when the acceleration is 68.7 in/s².

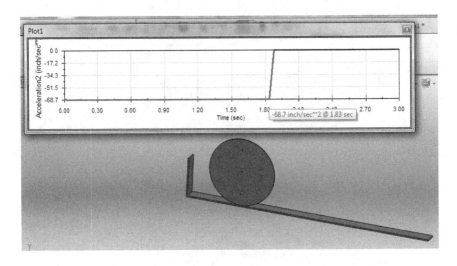

Fig. 14.142 Simulation results for Problem 14.7

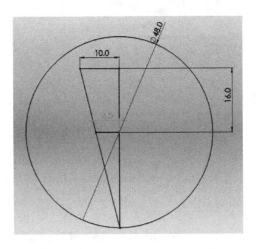

Fig. 14.143 Rolling wheel motion relations.

Verification of results with analytical method

Analytical method

1. Motion Analysis — when center of A travels with an acceleration a, mass B travels with an acceleration $(1 + 0.667)a = 1.667a$ (see Figure 14.143). When mass B drops, mass A rotates counter-clockwise.

From the triangles formed, if mass B travels 1 in, the center of the cylinder will travel 0.6 in. Therefore, we have the relationships:

$$S_c = 0.6S_B; \quad V_c = 0.6V_B$$
$$S_c = 0.6S_B; \quad V_c = 0.6V_B$$

Conversely, $S_B = 1.667S_C$; $V_B = 1.667V_C$

2. Figure 14.144 shows the FBDs.

3. Applying equilibrium conditions

For Cylinder:

$$\sum M_o = 0 : T(3.333) + I_C\alpha - (m_A a)r = 0$$

$$T(3.333) = 8.7\left(\frac{r}{2}\right) + \left(\frac{140}{32.2}\right)(a)(2)$$

$$T = \left(\frac{8.7}{3.333}\right)\left(\frac{r}{2}\right) + \left(\frac{140}{3.333 \times 32.2}\right)(2)(a) = 1.306a + 2.61a$$

$$\therefore T = 3.917a \tag{14.1}$$

For mass B:

$$\sum F_y = 0 : 25 - m_B a_B - T = 0$$

$$T = 25 - \left(\frac{25}{32.2}\right)\left(\frac{3.333}{2}a\right) = 25 - 1.29a \tag{14.2}$$

Using equations (14.1) and (14.2): $3.917a = 25 - 1.29a$, leading to:

$$(3.917 + 1.29)a = 25$$

$$a = \frac{25}{5.209}$$

$$a = 4.8\,\text{ft/s}^2$$

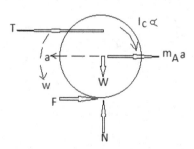

Fig. 14.144 FBD of force system.

The acceleration of B, $a_B = \frac{3.33}{2}a = 1.665(4.8) = 8\,\text{ft/s}^2$. Frictional force acting on cylinder A:

$$\sum F_x = 0 : m_A.a - T + F = 0$$

$$F = T - m_A.a = 3.917a - \frac{140}{32.2} \cdot a = 1.956a - 4.347a = -2.391a$$

$$\therefore\ F = -11.48\,\text{lb}$$

The displacement A travels in 3 seconds:

$$s = v_o t + \frac{1}{2}at^t = 0 + \frac{1}{2}(8)(3)^2 = 21.6\,\text{ft}$$

SolidWorks simulation

The result plot for Problem 14.7 shown in Figure 14.142 gives an acceleration of $68.7\,\text{in/s}^2$. Converted to the same units, the acceleration is $5.73\,\text{ft/s}^2$. Our simulation result is good enough when compared to the analytical method.

Problem 14.7

This problem relates to a governor used in speed regulation. This governor was used for the cover design of this textbook.

The governor shown in Figure 14.145 regulates the speed of a motor by having the weights shown, rotating on a vertical shaft. Collar C is fixed to

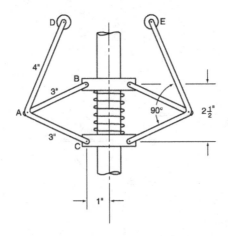

Fig. 14.145 Governor.

the shaft and collar B moves down due to centrifugal action of the weights. The weights at D and E are 0.5 lb each and the weight of the arms can be neglected. For the position shown at 400 rpm, determine the compressive load on the spring.

SolidWorks solution

Open New SolidWorks Part Document.
Sketch one-half of the arms (see Figure 14.146).
Create Plane1, Plane2, and Plane3 for the Sweep operations (see Figure 14.147).
Sketch a circle on Plane1 and Sweep it through the part of the arm (see Figure 14.148).
Repeat for a second circle on Plane2.
Sketch a circle on Plane3 and Sweep it through the part of the arm (see Figure 14.149).
Create a sphere (D or E).
Create the shaft.
Create collar B.
Create the spring.
Create a base.

Figure 14.150 shows how the arm is checked to ensure that it fits correctly to the collar C which is fixed to the shaft.

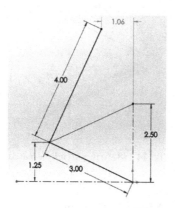

Fig. 14.146 One-half of arm description.

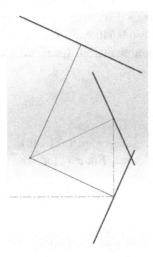

Fig. 14.147 Planes created.

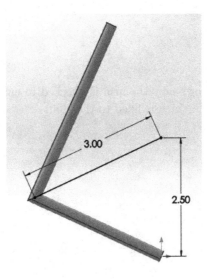

Fig. 14.148 Arm partially created.

Assembling the Governor:

Open New SolidWorks Assembly Document.

Insert the Base as the first part.

Insert all the parts created (see Figure 14.151 for the assembly).

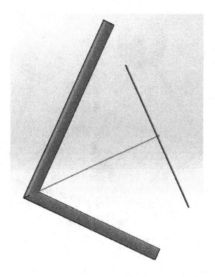

Fig. 14.149 Arm partially created.

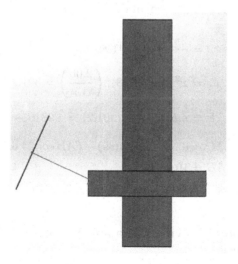

Fig. 14.150 Checking alignment.

Fig. 14.151 Governor for regulating speed of a motor.

Analysis

The FBD for one-half of the arm of the governor is shown in Figure 14.152.

$$\omega = \frac{2\pi(400)}{60} = 41.88 \,\text{rad/s}$$

$$a_r = \omega^2 r = (41.88)^2 \left(\frac{2.06}{12}\right) = 301.2 \,\text{ft/s}^2 \leftarrow$$

$$F = ma_r = \frac{0.5}{32.2}(301.2) = 4.68 \,\text{lb} \leftarrow$$

$$\sum M_c = 0; \quad (4.68)(4.89) - (AB \cos 24.62)(2.5) = 0$$

$$\therefore AB = 10.7 \,\text{lb}$$

$$\therefore Spring \, load = 2(10.7 \times \sin 24.62) = 8.92 \,\text{lb}$$

Manual Method of Solving Dynamics Problems using SolidWorks

This section presents a manual method of solving dynamic problems using SolidWorks as a 'calculator'. This calculator-based approach means that users have to be very familiar with the steps involved in solving the problem being investigated, and then SolidWorks tools are used to facilitate the process. Some examples are given to clarify the concept. The calculator-based

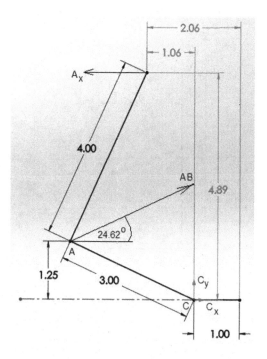

Fig. 14.152 FBD of one-half of the arm of the governor.

approach is not a substitute for the material presented in this chapter, which is based on the simulation power of SolidWorks.

Problem 14.8

Determine the acceleration of the 150 lb block in Problem 14.1 if the coefficient of kinetic friction is 0.4.

SolidWorks solution

OA is the vector representing the mass of 150 lb (see Figure 14.153).
AB is the vector representing the load of 90 lb downward at an angle of 45° to the horizontal.
BC is the vector representing the load of 60 lb upward at an angle of 60° to the horizontal.
OC is the resultant vector having a value of 217 lb with the horizontal component of 115.6 lb and vertical (normal) component of 183.64 lb. Due to friction, the frictional force is 0.4 (183.64) = 73.44 lb. Therefore, the

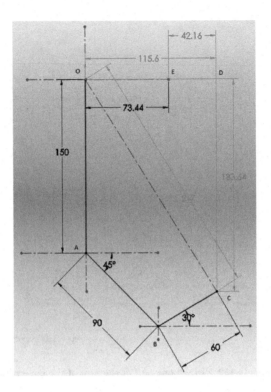

Fig. 14.153 Calculator for Problem 14.1.

effective horizontal load $= 115.6 - 73.44 = 42.16$ lb. The acceleration is given as:

$$a = \frac{42.16}{(150/32.2)} = 9.05 \, \text{ft/s}^2$$

Problem 14.9

Determine the acceleration of the 30 kg block in Problem 14.2; the coefficient of kinetic friction is 0.7.

SolidWorks solution

OA is the vector representing the upward force on the 25° inclined plane of 400 N.

BD is the vector representing the weight of 294.3 N (30 × 9.81) downward; the components along the inclined plane and perpendicular plane are 124.38 N and 266.73 N respectively.

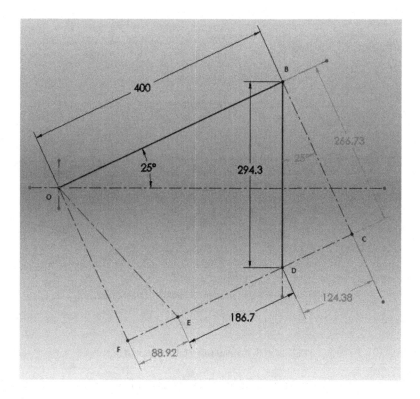

Fig. 14.154 Calculator for Problem 14.2.

Due to friction, the frictional force is 0.7 (266.73) = 186.7 N. Therefore, the effective load along the inclined plane = 400−(186.7+124.38) = 88.92 N. See Figure 14.154 for the solution.

The acceleration is given as:

$$a = \frac{88.92}{(30)} = 2.964 \, \text{m/s}^2$$

Problem 14.10

Determine the acceleration of block A down the slope in Problem 14.4 in this chapter; the coefficient of kinetic friction is 0.2.

SolidWorks solution

Choose a scale of 1:10.

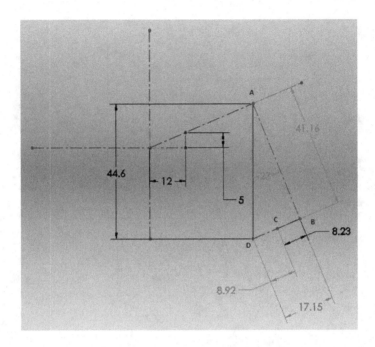

Fig. 14.155 Calculator for Problem 14.3.

AD is the vector representing the weight of 445.95 N (45 × 9.81) downward; the components along the inclined plane and perpendicular plane are 171.5 N and 411.6 N respectively. Using the scale, the values become 44.595 N (load), 17.15 N and 41.16 N respectively (see Figure 14.155). Due to friction, the frictional force is 0.2 (41.16) = 8.233 N. Therefore, the effective load along the inclined plane = (17.15 − 8.23) = 8.92 N. Applying the scale, the load = 10(8.92) = 89.2 N. The acceleration is given as:

$$a = \frac{89.2}{(45)} = 1.98 \, \text{m/s}^2$$

Summary

This chapter has discussed exhaustively SolidWorks approach to solving dynamics of bodies' problems for which the inertia method is used as a check for linear and plane motions. The SolidWorks concept is a new one which is an asset for teaching dynamics and for use in practice since the simulation approach is more intuitive and user-friendly than the theoretical method.

Exercises

P1. Determine the acceleration of the 150 lb block in Figure P1 if the coefficient of friction is 0.4.

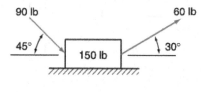

Fig. P1

P2. Resolve P1 with the following:

$$P_1 = 180\,lb; \quad P_2 = 120\,lb; \quad W = 300\,lb; \quad \theta_1 = 45°; \quad \theta_2 = 30°$$

P3. Determine the acceleration of the 50 lb block in Figure P2.

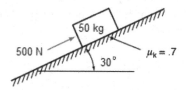

Fig. P2

P4. Resolve P3 with the following:

$$W = 150\,kg; \quad P = 1500\,N; \quad \theta = 30°.$$

P5. A 100 kg cart is accelerated horizontally by a $P = 300\,N$ force pulling at an angle of $\theta = 45°$ above horizontal. Neglecting rolling resistance, determine the acceleration of the cart.

P6. Determine the acceleration of block A in Figure P3.

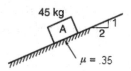

Fig. P3

P7. Determine the acceleration of the 200 lb block in Figure P4 if the force P = 40 lb.

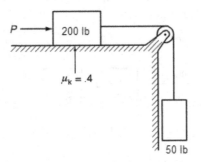

Fig. P4

P8. The coefficient of friction for mass B (100 kg) is $\mu = 0.25$ (kinetic). The rise and fall of the inclined plane are $y = 3$ and $x = 4$. Determine the acceleration of mass A, if it has a mass of (a) 30 kg and (b) 50 kg (see Figure P5).

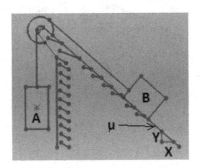

Fig. P5

P9. Cylinder A in Figure P6 has a diameter of 0.6 m and a mass of 260 kg. Assume that there is no slippage of A and that mass B is released from rest: Determine (a) the tension in the rope and (b) The distance that B will drop in 20 seconds. (Neglect the mass and inertia of the rope and pulley.)

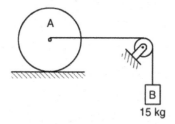

Fig. P6

Reference

Walker, K. M., *Applied Mechanics for Engineering Technologists*, 8th Edition, Prentice Hall, Upper Saddle River, NJ, 2007.

Chapter 15

Kinetics: Work and Energy

Objectives: When you complete this chapter you will have understanding on how to:

- Calculate the work of a constant force.
- Calculate the work of a variable force, such as spring force.
- Apply the conservation of energy principles to linear, angular, and plane motion.
- Apply SolidWorks Motion to solve problems related to the conservation of energy principles for linear, angular, and plane motions.

In this chapter, work and energy are covered for linear, rotational, and plane motions.

Work and Energy: Linear

Whenever a force is applied to an object, causing the object to move, work is being done by the force. If a force F is applied but the object does not move, no work is done; if a force is applied and the object moves a distance s in a direction other than the direction of the force, less work is done than if the object moves a distance s in the direction of the applied force.

Work $U = Fs$ (see Figure 15.1).
Where $U =$ Work, in N-m (Joules) or ft-lb.
$F =$ Force, N or lb.
$s =$ distance, m or ft.

Work of spring

The spring constant k is a measure of the stiffness of a spring. Large value of k implies a stiff spring, while small value of k implies a soft spring. To compress a spring by a distance x we must apply a force $F_{ext} = kx$. By Newton's third law, if we hold a spring in a compressed position, the spring

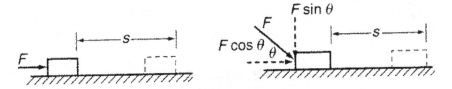

Fig. 15.1 Force and distance for definition of work.

exerts a force $F_s = -kx$. This is called a *linear restoring force* because the force is always in the *opposite direction from the displacement.* Work = (average force)(distance).

$U = \dfrac{1}{2}kx^2$ in N-m or ft-lb, where k = spring constant, N/m or lb/ft and

x = compression of extension of the spring.

Potential energy

When an object is dropped from a height, it picks up speed along the way as it falls down. This means there must be a net force on the object, doing work. This force is the force of gravity, with a magnitude equal to mg, the weight of the object. An alternate way of looking at this is to call this the gravitational potential energy. An object with potential energy has the potential to do work. In the case of gravitational potential energy, the object has the potential to do work because of where it is, at a certain height above the ground, or at least above something.

The work done by the force of gravity is the force multiplied by the distance, so if the object drops a distance h, gravity does work on the object equal to the force multiplied by the height lost, which is: work done by gravity. Mathematically, potential energy can be defined as:

$PE = mgh$ or Wh, in N-m or ft-lb, where m = mass in kg or slug,

g = acceleration due to gravity, m/s^2 or ft/s^2, and

h = vertical height with respect to a reference point, m or ft.

Kinetic energy

An object has kinetic energy if it has mass and if it is moving. It is energy associated with a moving object, in other words. For an object traveling at a speed v and with a mass m, the kinetic energy is given by:

$$KE = 1/2\, mv^2 \text{ with the unit in N-m or ft-lb,}$$
$$\text{where } v = \text{velocity m/s or ft/s.}$$

Conservation of energy: Angular

The law of conservation of energy states that energy cannot be created or destroyed instead it can merely be changed from one form of energy to another. Energy often ends up as heat, which is thermal energy (kinetic energy, really) of atoms and molecules. Kinetic friction, for example, generally turns energy into heat, and although we associate kinetic friction with energy loss, it really is just a way of transforming kinetic energy into thermal energy. The law of conservation of energy occurs always, everywhere, in any situation in life where there is transformation of energy from one form into another. The total energy of a system is always constant. Consider a block that slides from station 1 to station 2 as shown in Figure 15.2.

In station 1 we have potential energy while kinetic energy is 0.
In station 2 we have kinetic energy while the potential energy is 0.
So, ΔPE = ΔKE + friction work.

A force applied to a body that causes it to rotate creates torque. This torque acts to angularly accelerate a spinning object. The equation for

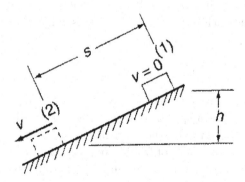

Fig. 15.2 A moving block that losses potential energy and gains potential energy.

angular kinetic energy is given as:

Angular $KE = 1/2I\omega^2$ in N-m or ft-lb, where I = mass moment of inertia in $kg - m^2$ or $slug - ft^2$ and ω = angular velocity in radian/s.

Conservation of energy: Plane Motion

In plane motion, both translation and rotation are present. Therefore, we apply all the principles so far to solve the problem of plane motion.

SolidWorks Solution Procedure for Work and Energy Problems

In this chapter, we lay out the SolidWorks solution procedure as follows:

1. **Model** the parts that define the problem being solved.
2. Create an **Assembly** of the parts and apply **Mating** conditions between the parts.
3. **Add-In** the **Motion Analysis** option.
4. Select **Motion Analysis** since Animation and Basic motion are not sufficient to solve the problem.
5. Create a **Spring** between parts as it should be in practice, using the features in **Motion Analysis** interface.
6. Define **Force** on the appropriate position.
7. Click the **Calculate** tool.
8. View the **Results**.

These steps are now applied to solve some problems using SolidWorks in order to show how this CAD software is not only used for design and FEA analysis of machines but for also for solving problems in applied mechanics.

Problem 15.1

The cylinder in Figure 15.3 has a mass of 25 kg, while the spring is at its free length and has a spring constant of 800 N/m. The diameter of the cylinder is D = 800 mm and other dimensions given are y_1 = 200 mm and y_2 = 200 mm. The coefficient of friction between the cylinder and the floor is 0.65. Assuming the system is initially at rest and there is no slipping, determine:

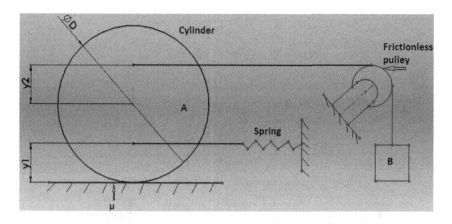

Fig. 15.3 A system of masses, pulley, and spring.

a. The mass of B; for this case assume B is lowered slowly and the distance travelled by B = 500 mm.
b. The velocity of B; for this case assume mass B = 30 kg, it is released suddenly and travels 300 mm.
c. The acceleration at the center of wheel A.

SolidWorks solution

We will do the following using SolidWorks:

1. Start a New SolidWorks Part document.
2. Model the cylinder and floor (body).
3. Start a New SolidWorks Assembly document.
4. Assemble the parts created.
5. Add-In a SolidWorks Motion Analysis module.
6. Replace the weight of block B hanging over the pulley, acting on the cylinder A with the force value of $25 \times 9.81 = 245.25$ N.
7. Create a spring and connect it between the cylinder and the body.
8. Simulate the dynamics of the systems realized.
9. Verify the solution using the analytical method.

Model building

(A) Body

1. Start a New SolidWorks Part document.

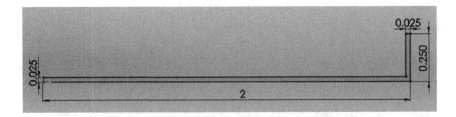

Fig. 15.4 Sketch for the body.

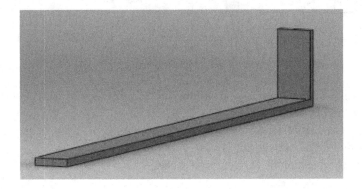

Fig. 15.5 Body.

2. Select the Front Plane.
3. Be in Sketch mode and sketch the profile shown in Figure 15.4.
4. Extrude 0.12 m (120 mm); the body is shown in Figure 15.5.
5. Save the body and exit.

(B) Cylinder

1. Start a New SolidWorks Part document.
2. Select the Front Plane.
3. Be in Sketch mode and sketch a circle of 0.8 m diameter.
4. Extrude 0.006313 m (6.313 mm); the cylinder is shown in Figure 15.6.
5. Apply Material: AISI 1015 Steel, Cold Drawn. (The value for extrusion is obtained after several trials to ensure mass of 25 kg.)
6. Click Mass Properties.
7. Save the body and exit.

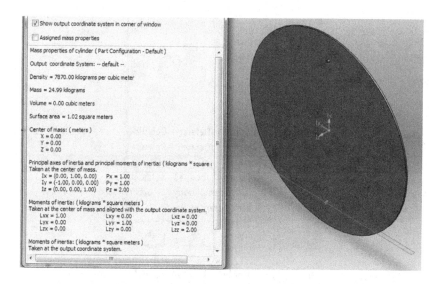

Fig. 15.6 Cylinder with mass properties shown.

Fig. 15.7 Assembly of cylinder and body.

Assembly modeling

1. Start a New SolidWorks Assembly document.
2. Select the base as the first part to insert.
3. Select the cylinder as the second part to insert (see Figure 15.7).

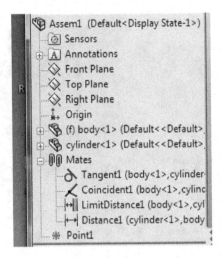

Fig. 15.8 FeatureManager showing mates.

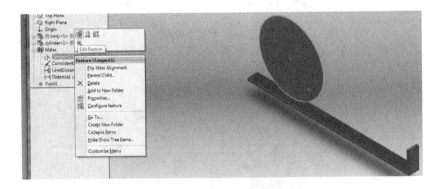

Fig. 15.9 Body and cylinder have a tangent mate.

4. Assign Mates (see Figure 15.8).
5. Tangent mate between body and cylinder (see Figure 15.9).
6. The distance of the pin on the cylinder and the top-face of the body is 0.2 m (see Figure 15.10).
7. The face of the pin (Plain1) on the cylinder and the Front Plane of the body are coincident (see Figure 15.11).
8. Limit distance between the center point (Point1) of the cylinder and right-hand side end-face of the body (Face<1>). The maximum distance is 1.5 m and the minimum is 1.20 m, giving a distance of 300 mm between the maximum and minimum range (see Figure 15.12).

Fig. 15.10 Distance mate between the pin on the cylinder and the top-face of the body.

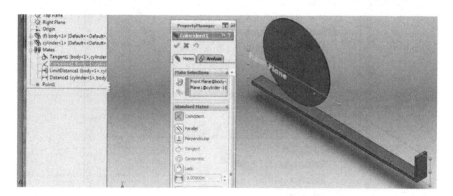

Fig. 15.11 Face of the cylinder pin (Plain1) and the Front Plane of the body are coincident.

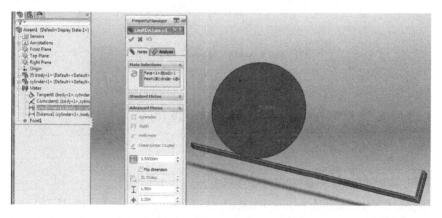

Fig. 15.12 Limit distance between the cylinder center point and right-hand face of the body.

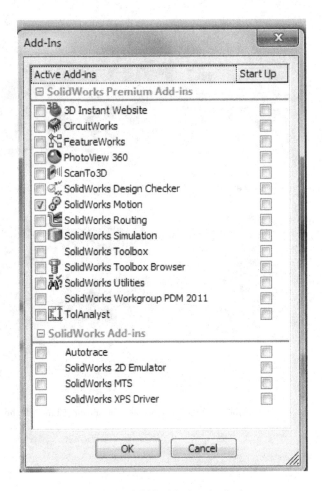

Fig. 15.13 Add-In Motion Analysis.

Select motion analysis

1. Add-In SolidWorkds Motion (see Figure 15.13).
2. Select Motion Analysis (see Figure 15.14).

Modeling of physical features: Spring

Physical features such as motors, spring, and contacts (see Figure 15.15) can be added to our assembly to define a dynamic system. Force and dampers can also be defined. In this example, a spring and force are defined.

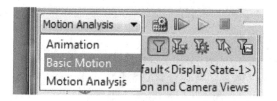

Fig. 15.14 Select Motion Analysis.

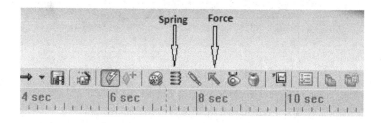

Fig. 15.15 Adding physical features to an assembly.

Define a spring

1. Click the **Spring** tool.
2. For the **Spring Parameters,** select the *inner face of the upright side of the body* as **Face <1>** the *pin on the cylinder* as **Face <2>** (see Figure 15.16).
3. For **Exponent of Spring Force Expression,** select **Linear**.
4. For **Spring Constant** set the value to be **800** (N/m).
5. For **Free Length** check **Update to model changes** (otherwise specify value).
6. Check **Damper** option.
7. For **Exponent of Damping Force Expression,** select **Linear**.
8. For **Damping Constant** set the value to be **20.0** (N(m/s)).
9. For **Coil Diameter** set the value to **0.01875 m**.
10. For **Number of Coils** set the value to **20**.

Define force

1. For **Direction** select **Action Only** (it is the default setting) (see Figure 15.17).
2. For **Action part and point of application of action** select the *upper pin on the cylinder.*

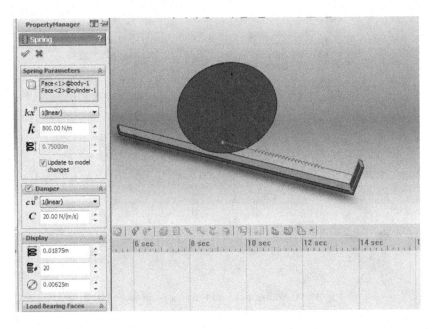

Fig. 15.16 Spring settings.

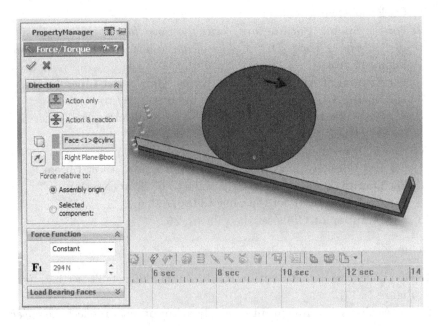

Fig. 15.17 Force settings.

3. For **Direction of Force** select the **Right Plane** of the body or cylinder.
4. For **Force Function** select **Constant** from the pull-down menu.
5. For **Constant Force value** type **294 N**.

Results

1. Click the **Calculate** tool.
2. View the **Results**.
3. Right-click **Plot1** under the **Results** folder and select the **Results** options (see Figure 15.18).

The velocity result (see Figure 15.19) shows that the velocity of the periphery of the cylinder picks up from 0 to a maximum of 4.1 m/s and drops to 0 at 0.1025 seconds.

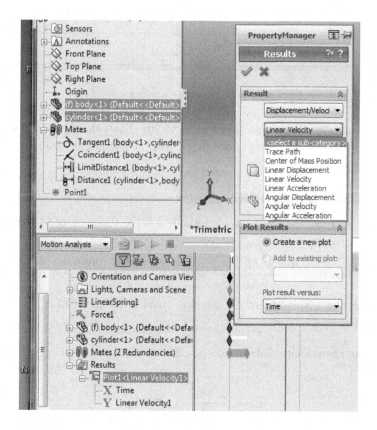

Fig. 15.18 Results options.

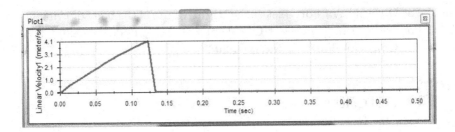

Fig. 15.19 Result for velocity.

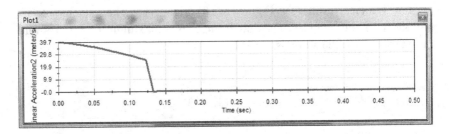

Fig. 15.20 Result for acceleration.

The acceleration result (see Figure 15.20) shows that the acceleration of the periphery of the cylinder drops from a maximum of 39.7 to 0 at 0.1025 seconds.

Verification of SolidWorks results

(a) We will use the 'calculator' method to solve this part of the problem (see Figure 15.21(a)).

$$\text{Mass B} = \frac{K \times s \times (r_1/r_2)}{9.81} = \frac{800 \times 0.1667 \times (200/600)}{9.81} = 4.53 \,\text{kg}$$

(b) The new displacement for 300 mm travel is 100 mm (0.1 m) as shown in Figure 15.21(b).

From the velocity distribution (see Figure 15.21(c)), $\frac{V_C}{0.4} = \frac{V_B}{0.6}$; $\therefore V_C = 0.667 V_B$

$$\omega = \frac{2/3 \, V_B}{0.4} = 1.667 \, V_B$$

Moment of inertia: $I_C = 1/2 m_A r_A^2 = 1/2 (25)(0.4)^2 = 2 \,\text{kg.m}^2$.

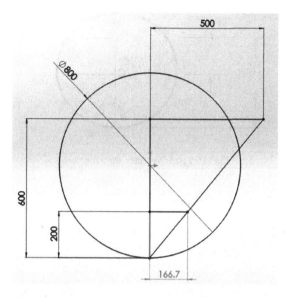

Fig. 15.21(a) Velocity distribution.

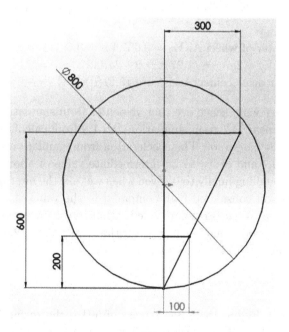

Fig. 15.21(b) Displacement distribution.

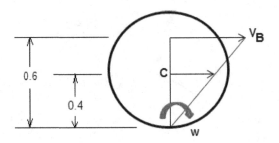

Fig. 15.21(c) Velocity distribution.

Conservation of energy equation:

$-PE_B + \Delta KE_B + \Delta KE_A + \Delta KE_{\omega A} + \Delta V_S = 0; \quad m_B = 30\,kg; \quad h = 0.3\,m$

$(30 \times 9.81 \times 0.3) = 1/2(30)V_B^2 + 1/2(25)(0.667\,V_B)^2 + 1/2(2)(1.667\,V_B)^2$
$$+ 1/2(800)(0.1)^2$$

$88.29 = 15\,V_B^2 + 5.56\,V_B^2 + 2.778\,V_B^2 + 4; \quad \Rightarrow 23.338\,V_B^2 = 84.29$

$\therefore\ V_B = \sqrt{\dfrac{84.29}{23.338}} = 1.9\,m/s \downarrow$

(c) At the center of wheel A, $V_C = 0.667\,V_B = 0.667(1.9) = 1.27\,m/s$
But $V_c^2 = V_0^2 + 2ah; \Rightarrow (1.27)^2 = 0 + 2a(0.2); \Rightarrow a = \frac{1.607}{2(0.2)} = 4.02\,m/s^2$
(See center displacement in Figure 15.21(d).)

The conclusions reached are that velocities from simulation and calculation are not near (4.1 from simulation and 1.9 from calculation) and the
accelerations deviate a lot. The acceleration from simulation is noisy as it
starts from 39.7 and drops to an intermediate value of about 25 in 0.125
seconds before falling finally to 0. Even when we take the intermediate value
of 25, this is still considered high compared to the value of 4.02 obtained
from calculation. What is recommended is that the average should be taken
for simulation results: that is 2.05 m/s for the velocity and 12.5 m/s^2 for the
acceleration at the 0.125–0.128 seconds band.

Problem 15.2

The cylinder in Figure 15.22 has a mass of 50 kg, the spring is at its free
length and has a spring constant of 600 N/m. Assuming the system is
initially at rest and there is no slipping, determine:

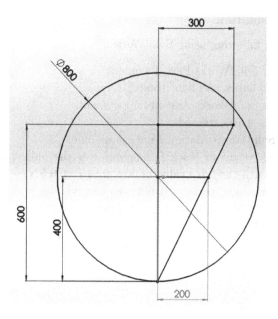

Fig. 15.21(d) Displacement distribution.

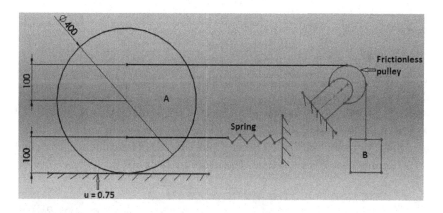

Fig. 15.22 A system of masses, pulley, and spring.

a. The mass of B; for this case assume B is lowered slowly and the distance travelled by B = 250 mm.

b. The velocity of B; for this case assume mass B = 40 kg, it is released suddenly and travels 400 mm.

c. The acceleration at the center of wheel A.

SolidWorks solution

We will do the following using SolidWorks:

1. Start a New SolidWorks Part document.
2. Model the cylinder and floor (body).
3. Start a New SolidWorks Assembly document.
4. Assemble the parts created.
5. Add-In a SolidWorks Motion Analysis module.
6. Replace the weight of block B hanging over the pulley, acting on the cylinder A with the force value of $50 \times 9.81 = 490.5$ N.
7. Create a spring and connect it between the cylinder and the body.
8. Simulate the dynamics of the systems realized.
9. Verify the solution using analytical method.

Model building

(A) Body

1. Start a New SolidWorks Part document.
2. Select the Front Plane.
3. Be in Sketch mode and sketch the profile shown in Figure 15.23.
4. Extrude 0.12 m (120 mm); the body is shown in Figure 15.24.
5. Save the body and exit.

(B) Cylinder

1. Start a New SolidWorks Part document.
2. Select the Front Plane.
3. Be in Sketch mode and sketch a circle of 0.4 m diameter.
4. Extrude 0.050504 m (50.504 mm); the cylinder is shown in Figure 15.25.
5. Apply Material: AISI 1015 Steel, Cold Drawn. (The value for extrusion is obtained after several trials to ensure mass of 50 kg.)

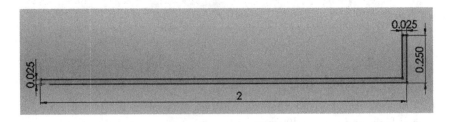

Fig. 15.23 Sketch for the body.

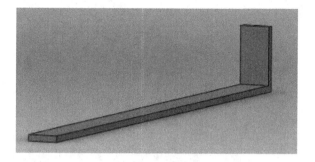

Fig. 15.24 Body.

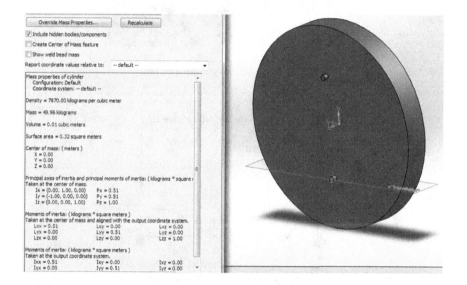

Fig. 15.25 Cylinder with mass properties shown.

6. Click Mass Properties.
7. Save the body and exit.

Assembly modeling

1. Start a New SolidWorks Assembly document.
2. Select the base as the first part to insert.
3. Select the cylinder as the second part to insert (see Figure 15.26).
4. Assign Mates (see Figure 15.27).

Fig. 15.26 Assembly of cylinder and body.

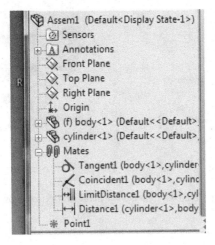

Fig. 15.27 FeatureManager showing mates.

5. Tangent mate between body and cylinder (see Figure 15.28).
6. The distance of the pin on the cylinder and the top-face of the body is 0.1 m (see Figure 15.29).
7. The face of the pin (Plain1) on the cylinder and the Front Plane of the body are coincident (see Figure 15.30).

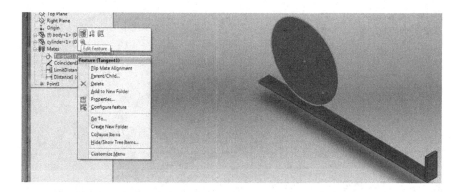

Fig. 15.28 Body and cylinder have a tangent mate.

Fig. 15.29 Distance mate between the pin on the cylinder and the top-face of the body.

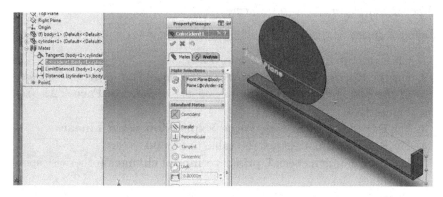

Fig. 15.30 Face of the cylinder pin (Plain1) and the Front Plane of the body are coincident.

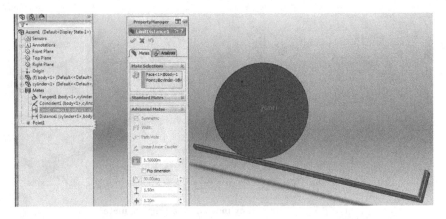

Fig. 15.31 Limit distance between the cylinder center point and right-hand face of the body.

8. Limit distance between the center point (Point1) of the cylinder and right-hand side end-face of the body (Face <1>). The maximum distance is 1.5 m and the minimum is 1.20 m, giving a distance of 300 mm between the maximum and minimum range (see Figure 15.31).

Select motion analysis

1. Add-In SolidWorkds Motion (see Figure 15.32).
2. Select Motion Analysis (see Figure 15.33).

Modeling of physical features: Spring

Define a spring

1. Click the **Spring** tool (see Figure 15.34).
2. For the **Spring Parameters**, select the *inner face of the upright side of the body* as **Face <1>** the *pin on the cylinder* as **Face <2>** (see Figure 15.35).
3. For **Exponent of Spring Force Expression**, select **Linear**.
4. For **Spring Constant** set the value to be **600** (N/m).
5. For **Free Length** check **Update to model changes** (otherwise specify value).
6. Check **Damper** option.
7. For **Exponent of Damping Force Expression**, select **Linear**.
8. For **Damping Constant** set the value to be **20.0** (N(m/s)).

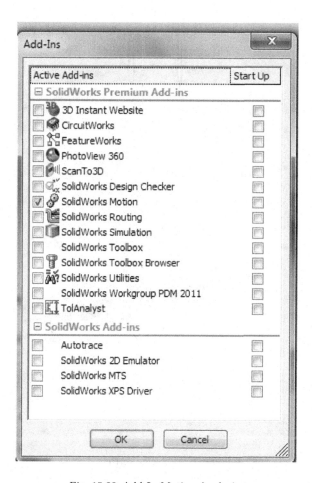

Fig. 15.32 Add-In Motion Analysis.

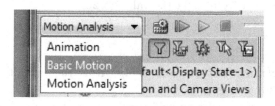

Fig. 15.33 Select Motion Analysis.

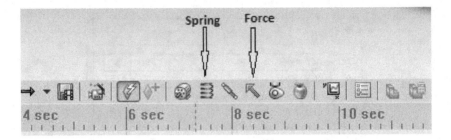

Fig. 15.34 Adding physical features to an assembly.

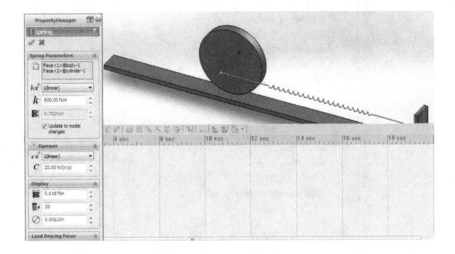

Fig. 15.35 Spring settings.

9. For **Coil Diameter** set the value to **0.01875 m**.
10. For **Number of Coils** set the value to **20**.

Define force

1. For **Direction** select **Action Only** (it is the default setting) (see Figure 15.36).
2. For **Action part and point of application of action** select the *upper pin on the cylinder*.
3. For **Direction of Force** select the **Right Plane** of the body or cylinder.
4. For **Force Function** select **Constant** from the pull-down menu.
5. For **Constant Force value** type **490.5 N**.

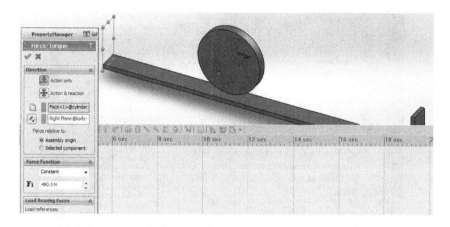

Fig. 15.36 Force settings.

Results

1. Click the **Calculate** tool.
2. View the **Results**.
3. Right-click **Plot1** under the **Results** folder and select the **Results** options (see Figure 15.37).

The velocity result (see Figure 15.38) shows that the velocity of the periphery of the cylinders pick up from 0 to a maximum of 2.8 m/s and drops to 0 at 0.18 seconds.

The acceleration result (see Figure 15.39) shows that the acceleration of the periphery of the cylinders drops from a maximum of 16.9 to 0 at 0.18 seconds.

Verification of SolidWorks results

(a) We will use the 'calculator' method to solve this part of the problem (see Figure 15.40(a)).
$$\text{Mass B} = \frac{K \times s \times (r_1/r_2)}{9.81} = \frac{600 \times 0.0833 \times (100/300)}{9.81} = 1.7 \, \text{kg}$$
(b) The new displacement for 400 mm travel is 133.3 mm (0.133 m) as shown in Figure 15.40(b).

From the velocity distribution (see Figure 15.40(c)), $\frac{V_C}{0.2} = \frac{V_B}{0.3}$; $\therefore V_C = 0.667 V_B$

$$\omega = \frac{2/3 V_B}{r} = \frac{2/3 V_B}{0.2} = 3.33 \, V_B$$

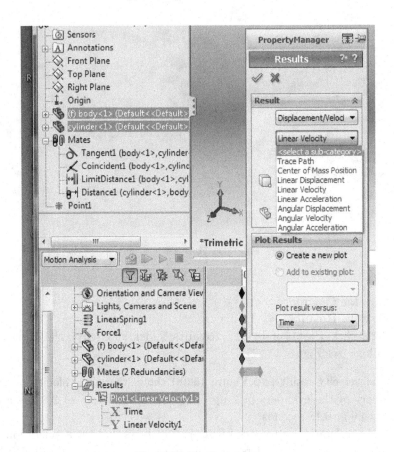

Fig. 15.37 Results options.

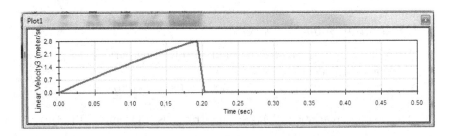

Fig. 15.38 Result for velocity.

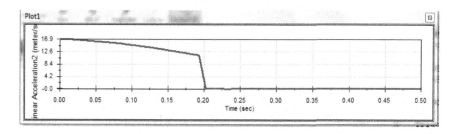

Fig. 15.39 Result for acceleration.

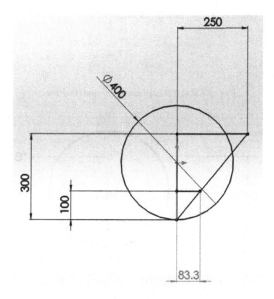

Fig. 15.40(a) Velocity distribution.

Moment of inertia: $I_C = 1/2 m_A r_A^2 = 1/2(50)(0.2)^2 = 1\,\text{kg.m}^2$

Conservation of energy equation:

$-PE_B + \Delta KE_B + \Delta KE_A + \Delta KE_{\omega A} + \Delta V_S = 0; \quad m_B = 40\,kg; \quad h = 0.4\,\text{m}$

$(40 \times 9.81 \times 0.4) = 1/2(40)V_B^2 + 1/2(50)(0.667\,V_B)^2 + 1/2(1)(3.337\,V_B)^2$
$\qquad\qquad\qquad + 1/2(600)(0.133)^2$

$156.96 = 20\,V_B^2 + 11.12\,V_B^2 + 5.56\,V_B^2 + 5.31; \quad \Rightarrow 36.66\,V_B^2 = 151.65$

$\therefore\; V_B = \sqrt{\dfrac{151.65}{36.66}} = 2.03\,\text{m/s}\;\downarrow$

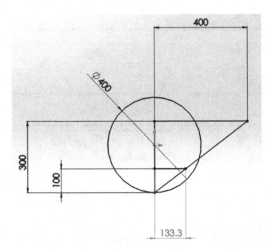

Fig. 15.40(b) Displacement distribution.

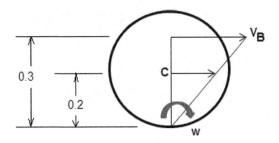

Fig. 15.40(c) Velocity distribution.

(c) At the center of wheel A, $V_C = 0.667 V_B = 0.667(2.03) = 1.35\,\mathrm{m/s}$
But $V_c^2 = V_0^2 + 2ah$; $\Rightarrow (1.35)^2 = 0 + 2a(0.263)$; $\Rightarrow a = \frac{1.84}{2(0.263)} =$
$3.45\,\mathrm{m/s^2}$. (See center displacement in Figure 15.40(d).)

The conclusions reached are that velocities from simulation and calculation are near (2.8 from simulation and 2.03 from calculation) but the
accelerations deviate greatly. The acceleration from simulation is noisy as
it starts from 16.9 and drops to an intermediate value of about 11 in 0.18 s
before falling finally to 0. Even when we take the intermediate value of 11,
this is still considered high compared to the value of 3.45 obtained from

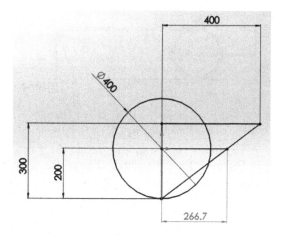

Fig. 15.40(d) Displacement distribution.

calculation. What is recommended is that the average should be taken for simulation results: that is 1.4 m/s for the velocity and 5.5 m/s² for the acceleration at the 0.18–0.21 seconds band.

Summary

This chapter has discussed the SolidWorks approach to solving dynamics of bodies' problems for which work and energy method is used as a check for linear and plane motions. The SolidWorks concept is a new one which is an asset for teaching dynamics and for use in practice since the simulation approach is more intuitive and user-friendly than the theoretical method.

Exercises

P1. Cylinder A in Figure P1 has a mass of 20 kg ($I_c = 0.9$ kg.m²). The spring has a free length of 0.5 m and a spring constant of 600 N/m. Assuming no slipping and neglecting the inertia of the cable and pulley, determine (a) the distance B will drop if lowered slowly and (b) the velocity of B if it is released from the position shown and has dropped 0.4 m.

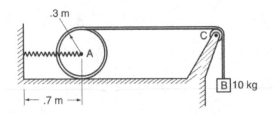

Fig. P1

P2. Cylinder A in Figure P2 weighs 322 lb. The spring is at its free length and has a spring constant of 10 lb/ft. If the system is initially at rest, determine the velocity of B after it has dropped 6 ft.

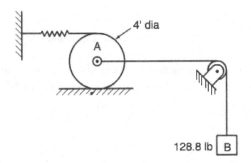

Fig. P2

P3. Cylinder A in Figure P3 weighs 300 lb while block B weighs 150 lb. The spring is at its free length and has a spring constant of 3 lb/ft. The coefficient of friction between the cylinder and the slope is 0.25. If the system is released from rest, determine the velocity of B after A has rolled 2 ft on the slope.

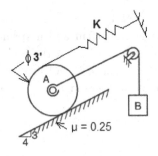

Fig. P3 Problem description.

P4. Cylinder A in Figure P4 has a mass of 260 kg and a diameter of 0.6 m. Assuming that there is no slippage of A and that mass B is released from rest, determine the velocity of B after A has rolled 0.3 m to the right.

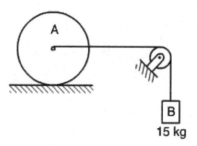

Fig. P4

Reference

Walker, K. M., *Applied Mechanics for Engineering Technologists*, 8th Edition, Prentice Hall, Upper Saddle River, NJ, 2007.

Chapter 16

Kinetics: Impulse and Momentum

Objectives: When you complete this chapter you will have understanding on how to:

- Calculate the impulse of moving objects using SolidWorks Motion tools.
- Calculate the momentum of moving objects using SolidWorks Motion tools.

What is Impulse?

When an object is subjected to a force for a very short time it is said to have received an *impulse*.

Impulse is given by force × time.

Impulse = $F \cdot t$, where F is in N and t is seconds and impulse is N-s.

In the imperial system, F is in lb, t is in seconds and impulse is lb-s.

What is Momentum?

When a mass moves with a velocity it generates *momentum*.

Momentum is given by mass × velocity, $M = mv$, where m is the mass and v is the velocity.

In the metric system m is in kg, v is m/s, and M is in kg-m/s or N-s (kg = N-s^2/m).

In the imperial system m is in slug (lb-s^2/ft), v is in f.t/s, and M is in lb-s.

Since velocity is a vector quantity, momentum is also a vector quantity.

Relation Between Impulse and Momentum

From Newton's second law F = ma, where m is the mass and a is the acceleration.

Acceleration: $a = v/t$, so $F = mv/t$ or $F \cdot t = mv$.

$F \cdot t$ is the impulse and mv is the momentum.

So the change in momentum of an object is equal to the impulse applied to the system.

Conservation of Momentum

A moving object may transfer or lose some of its momentum to another object. Consider a fast moving object that strikes and becomes attached to a slow moving object traveling in the same direction (see Figure 16.1).

The fast moving object exerts a force or impulse on the slow moving object, thereby speeding it up. The second object at that instant exerts an equal and opposite reaction or impulse on the first object, thereby slowing it down. While each object has experienced a change in velocity and therefore a change in momentum, the total momentum of the system remains the same.

This is known as *conservation of momentum.*

Referring to Figure 16.1, we see that conservation of linear momentum can be written as:

$$\text{Initial momentum} = \text{Final momentum}$$

$$(m_A v_A)_1 + (m_B v_B)_1 = (m_A v_A)_2 + (m_B v_B)_2,$$

where 1 is the initial stage and 2 is the final stage. We note that this is a vector addition, where direction is very important.

Impulse and Momentum using SolidWorks

From the definition of momentum units, momentum can be expressed as *force-second.* This is the format that we will use in implementing momentum using SolidWorks.

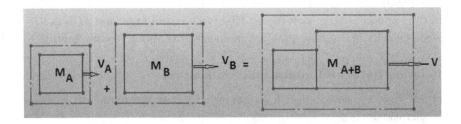

Fig. 16.1 Concept of momentum.

Problem 16.1

A sphere moves with applied force of 1.957 lb towards another sphere which is also moving with an applied force of 0.6854 lb. When they both collide and stick together what is their combined momentum in 0.08 seconds?

SolidWorks solution

Modeling of parts

1. Create a sphere R1.0 as shown in Figure 16.2.

Fig. 16.2 Sphere.

2. Create a base measuring $10 \times 15 \times 1/4$ (Figure 16.3).

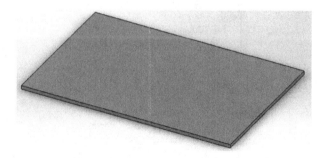

Fig. 16.3 Base.

Assembly modeling

1. Create an assembly of base and spheres (Figure 16.4).

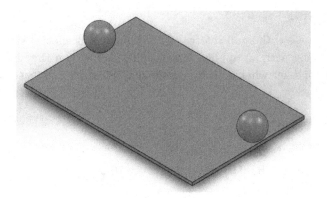

Fig. 16.4 Assembly of base and spheres.

Mates

1. Each sphere is tangential to the top surface of the base (Figure 16.5).
2. Each sphere is 5 in (at the midpoint) from the edge, opposite each other as shown. This arrangement ensures that the two spheres will collide during motion.

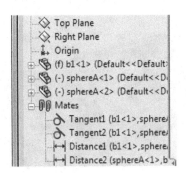

Fig. 16.5 Mates for spheres.

Motion analysis

The motion analysis tools in Figure 16.6 are in the MotionManager and are used to create motions.

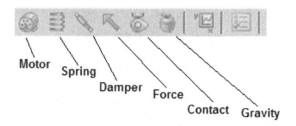

Fig. 16.6 Motion analysis tools.

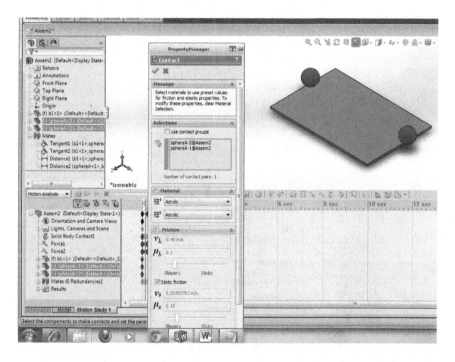

Fig. 16.7 Contact PropertyManager.

Create contact between the two spheres

1. From the PropertyManager of the Motion Analysis study (Figure 16.7), select the **two spheres** to be in contact during motion.
2. Click the **Contact** tool ⬚.
3. For **Material**, select **Acrylic** for both spheres (Figure 16.8).
4. For **Elastic Properties**, select **Restitution coefficient**.

Fig. 16.8 Setups for Contact PropertyManager.

Apply force to each of the two spheres

1. Click the **Force** tool and select each of the **two spheres** to apply force during motion.
2. Select the *left sphere* (see Figure 16.9), use the edge of the base as direction and apply a force of **1.957 lb**.
3. Click **OK** to accept.

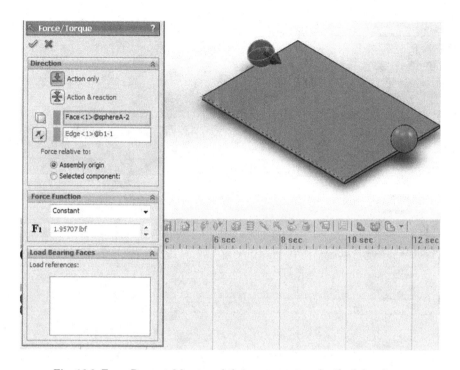

Fig. 16.9 Force PropertyManager defining parameters for the left sphere.

4. Select the *right sphere* (see Figure 16.10), use the edge of the base as direction and apply a force of **0.685 lb** (change direction if required).
5. Click **OK** to accept.

The Motion Analyis Manager is shown in Figure 16.11, containing the key properties or parameters defined. Set the end-time of the motion to **0.24** seconds (using the Key Properties of the time frame for the Assem2).

Motion simulation

1. Click **Calculate** to set the two sphere in motion.
   The two spheres are seen to be stuck together after some time and remain stuck till the end of the motion (see Figure 16.12).
2. Click **Results/Plot1 <Translational Momentum>** to give the plot shown in Figure 16.13.

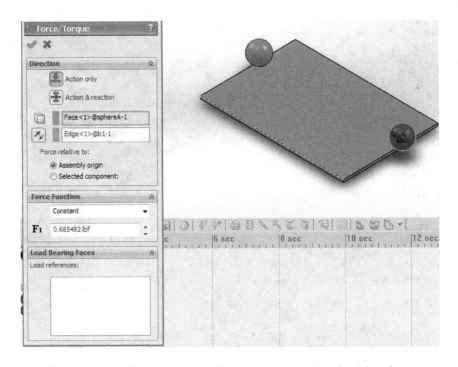

Fig. 16.10 Force PropertyManager defining parameters for the right sphere.

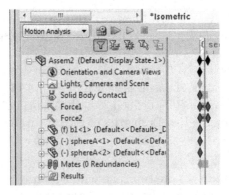

Fig. 16.11 Key Properties of the time frame for the Assem2.

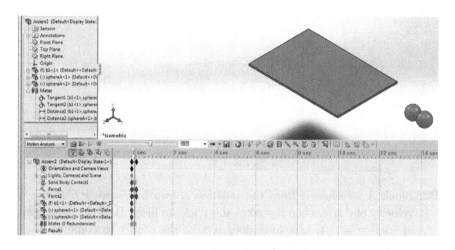

Fig. 16.12 Motion analysis.

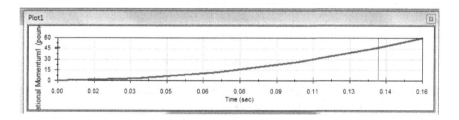

Fig. 16.13 Translational momentum plot with time.

Summary

This chapter described how SolidWorks is used to solve impact and momentum problems in applied mechanics. Solution to translational momentum problem were described, and rotational momentum problems can be similarly solved.

Exercises

P1. A plan view of objects A and B is shown in Figure P1. When the objects collide, they remain in contact. Determine the resulting velocity of A and B.

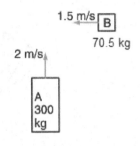

Fig. P1

P2. Block A is released from the position shown in Figure P2. It has a velocity of 6 m/s when it strikes and sticks to block B. If the coefficient of friction is 0.3, how long does it take the blocks to come to rest? What is the maximum velocity of B?

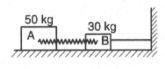

Fig. P2

P3. Object A and B in Figure P3 collide and remain in contact. Neglecting friction, determine the resultant velocity.

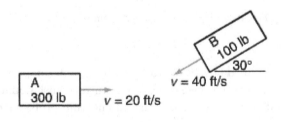

Fig. P3

Reference

Walker, K. M., *Applied Mechanics for Engineering Technologists*, 8th Edition, Prentice Hall, Upper Saddle River, NJ, 2007.

Appendix

Resources for Applied Mechanics Journals

Canadian Journal on Mechanical Sciences & Engineering
http://www.ampublisher.com/mse.html

International Journal of Applied Mechanics (IJAM)
http://www.worldscinet.com/ijam/

Journal of Applied Mechanics
http://journaltool.asme.org/Content/JournalDescriptions.cfm?journalId=1

Journal of Mechanical Design
http://www.asmedl.org/MechanicalDesign

Machining Science and Technology
http://www.tandf.co.uk/journals/titles/10910344.asp

Materials and Manufacturing Processes
http://www.tandf.co.uk/journals/titles/10426914.asp

Mechanics Based Design of Structures and Machines
http://www.tandf.co.uk/journals/titles/15397734.asp

Mechanics of Advanced Materials and Structures
http://www.tandf.co.uk/journals/titles/15376494.asp

Recent Advances in Applied and Theoretical Mechanics
http://www.wseas.us/e-library/conferences/2009/tenerife/MECHANICS/
MECHANICS-25.pdf

Transactions of the Canadian Society for Mechanical Engineering
http://www.tcsme.org/SubmissionGuidelines.html

Index

Printed in the United States
By Bookmasters